Lens Design Basics

Optical design problem-solving in theory and practice

IOP Series in Emerging Technologies in Optics and Photonics

Series Editor

R Barry Johnson a Senior Research Professor at Alabama A&M University, has been involved for over 50 years in lens design, optical systems design, electro-optical systems engineering, and photonics. He has been a faculty member at three academic institutions engaged in optics education and research, employed by a number of companies, and provided consulting services.

Dr Johnson is an IOP Fellow, SPIE Fellow and Life Member, OSA Fellow, and was the 1987 President of SPIE. He serves on the editorial board of Infrared Physics & Technology and Advances in Optical Technologies. Dr Johnson has been awarded many patents, has published numerous papers and several books and book chapters, and was awarded the 2012 OSA/SPIE Joseph W Goodman Book Writing Award for Lens Design Fundamentals, Second Edition. He is a perennial co-chair of the annual SPIE Current Developments in Lens Design and Optical Engineering Conference.

Foreword

Until the 1960s, the field of optics was primarily concentrated in the classical areas of photography, cameras, binoculars, telescopes, spectrometers, colorimeters, radiometers, etc. In the late 1960s, optics began to blossom with the advent of new types of infrared detectors, liquid crystal displays (LCD), light emitting diodes (LED), charge coupled devices (CCD), lasers, holography, fiber optics, new optical materials, advances in optical and mechanical fabrication, new optical design programs, and many more technologies. With the development of the LED, LCD, CCD and other electo-optical devices, the term 'photonics' came into vogue in the 1980s to describe the science of using light in development of new technologies and the performance of a myriad of applications. Today, optics and photonics are truly pervasive throughout society and new technologies are continuing to emerge. The objective of this series is to provide students, researchers, and those who enjoy self-teaching with a wide-ranging collection of books that each focus on a relevant topic in technologies and application of optics and photonics. These books will provide knowledge to prepare the reader to be better able to participate in these exciting areas now and in the future. The title of this series is Emerging Technologies in Optics and Photonics where 'emerging' is taken to mean 'coming into existence,' 'coming into maturity,' and 'coming into prominence.' IOP Publishing and I hope that you find this Series of significant value to you and your career.

Lens Design Basics

Optical design problem-solving in theory and practice

Christoph Gerhard

Laboratory of Analytical Measurement Technology, University of Applied Sciences and Arts, Göttingen, Germany

IOP Publishing, Bristol, UK, Bristol, UK

ISBN   978-0-7503-2240-9 (ebook)
ISBN   978-0-7503-2238-6 (print)
ISBN   978-0-7503-2241-6 (myPrint)
ISBN   978-0-7503-2239-3 (mobi)

DOI   10.1088/978-0-7503-2240-9

Version: 20201201

IOP ebooks

British Library Cataloguing-in-Publication Data: A catalogue record for this book is available from the British Library.

Published by IOP Publishing, wholly owned by The Institute of Physics, London

IOP Publishing, Temple Circus, Temple Way, Bristol, BS1 6HG, UK

US Office: IOP Publishing, Inc., 190 North Independence Mall West, Suite 601, Philadelphia, PA 19106, USA

This book is dedicated to my dear friend Dr Geoff Adams, the inventor of the optical design software WinLens, and to my family.

Contents

Foreword

This book is both humble and honest. It does not profess to offer a profound theoretical understanding of optics or even the subtle depths of wave theory or ray tracing. It does not offer a detailed analysis of the history and properties of different lens types such as the double Gauss or the microscope or the intricacies of zoom lens design.

Indeed optical design, as a topic, is no longer at the cutting edge of the field of computing algorithms, but it is every bit as vital to students and to engineers and scientists of many stripes. Some—a lucky few—will make optical design their career. But for many, optical design is something that will be part of their course of studies or will impinge on their everyday work—sometimes quite suddenly and dramatically. Consider the field of machine vision. Most engineers concentrate on processing the image collected by the CCD, but a poor choice of lens can mean that there is no image worthy of the name to process! However, a little optical knowledge would ensure that a clear crisp image, of the correct size and quality, is formed on the sensor.

The first part of the book provides an overview of optical design—'*the very basic equations and theory*', '*the common analysis tools and how to interpret them*', '*the practical issues that you will encounter*', '*the process from back of the envelope to the real working product*'. The second part of the book is an extensive series of worked exercises that are intended to be followed sitting in front of a PC. The exercises are sequenced to follow the flow of the preceding seven chapters, and to illustrate key points. Solutions are provided as an integral part of each question. This book uses the free WinLens3D software, but can easily be followed using your favourite lens design software.

The author, Prof Christoph Gerhard, has for many years given 'hands-on' PC based workshops introducing lens design. He has a rare ability to clearly communicate complex ideas. He knows by long experience the sort of questions that students and engineers have, and the joy when they suddenly see and understand what was previously opaque. This book is the fruit of that experience.

I have had the real pleasure of knowing and working with Christoph for nearly 15 years and count him as one of my true friends. We have worked together on many conference papers and stood on the stands together at many a technical exhibition. He has had a significant impact on the development of WinLens3D. The feedback from his courses around Europe has provided the spur for adding new features or improving existing ones so that they are more comprehensive and/or easier to use—priceless information for a tool maker.

Shoreham-by-Sea, September 2020

Dr Geoff Adams

How to use this book

In the first seven chapters, basics of optical system design are presented. The appropriate theory is discussed and illustrated, and practical approaches to solving particular imaging tasks are provided by several 'Case examples'. A number of comprehension questions moreover help to reflect on the basic phenomena of optical imaging. Info boxes, titled 'Focus on...[1]' are embedded in the text and point to exercises in chapter eight where the theory can be applied by hands-on exercises. At the end of the first seven chapters, the essential key points are summarised in a bottom line. A list of references further provides an overview on additional literature.

The eighth and last chapter contains a series of exercises—sorted by subject areas or topics—that represent typical tasks in lens design and optical system design. Different aspects, i.e. the definition of start systems, the creation, evaluation and optimisation of lenses and systems and the simulation of manufacturing errors and tolerances are covered. The detailed step-by-step solutions to the particular exercises including additional information are provided in the appendix of this book. The software used for solving the exercises is introduced and additional information and resources are given at the beginning of chapter eight.

Finally, the appendix contains a formulary of the most important lens design equations as well as a list of recommended literature.

---

[1] ...in the truest sense of the word...

Acknowledgments

This book is based on the script of the lecture "Computer-Assisted Optical System Design" held by the author at the Polytechnic University of Milan, Italy from 2016–2018. The author thus thanks Professor Gianluca Valentini from this university for making the lecture possible and for his extraordinary help and support! Further, the author thanks Professor Giuseppe della Valle for his help during the organisation of the lectures. Special thanks are also expressed to all the students who attended the lecture and who contributed to the successive improvement of the lecture script and —as a consequence—the present book by extensive and fruitful discussion.

The author further thanks Dr Geoff Adams, the inventor of the software used in this work, for his friendship. Without his input, support and effort it would have been impossible to write the present book. And of course, such a book project is only accomplished by the support and help of the editor, namely Ashley Gasque and her team, especially Robert Trevelyan.

Finally, very special thanks are also expressed to Martina, Benjamin and Philipp. Thank you for being with me!

Ebergötzen, in September 2020
Christoph Gerhard

Author biography

Christoph Gerhard

Christoph Gerhard is a professor for Physics and Instrumental Analytics and head of the Laboratory of Analytical Measurement Technology at the University of Applied Sciences and Arts in Göttingen, Germany.

After receiving his general qualification for university entrance, he passed an apprenticeship as an optics technician and subsequently worked as skilled technician and instructor. He then completed his diploma study of Precision Optics Manufacturing Technology as well as a master study of Optical Engineering and Photonics in Göttingen, Orsay and Bremen. After professional activity as a research associate and scientific coordinator, he earned a doctoral degree in Physical Technologies at Clausthal University of Technology in 2014. Following a Postdoc period with the Fraunhofer Society, he held the position of an adjunct professor for Physical Technologies in Göttingen and additionally was a visiting professor for Optical System Design in the frame of the MSc Programme Engineering Physics at the Polytechnic University of Milan. In 2017, he became a full professor for Laser and Plasma Technology at the Technical University of Applied Sciences of Wildau where he also was the coordinator responsible for the MEng Programme Photonics. Moreover, he has been a lecturer for optical system design at the Laser Academy, Hanover, since 2015.

Christoph Gerhard has published more than 50 articles on optics and optical system design. The present work is his fourth textbook in the field of optics.

Chapter 1

Fundamentals

The basis of any optical system and thus lens design is the effect of refraction at optical interfaces. Since this effect is wavelength-dependent, dispersion of optical media is another important issue. This also applies to reflection as well as intensity-related phenomena that result from the wave character of light. For instance, diffraction is an essential point for the definition of an imaging system's resolution and interference represents the basis of dielectric coatings. For an understanding of classical aberrations and other defects of imaging systems the knowledge of these phenomena is crucial. The most important fundamentals are thus introduced in this chapter.

1.1 Light waves and light rays

In optical system design, light is usually described as a bundle of **light rays** in most cases. Those light rays are given by the normal on the wave front emitted by a light source as, for example, the Sun. In some rare cases, for example if sunlight shines through gaps in clouds or crest coronets, such sunrays become visible in nature.

As shown in figure 1.1, a light ray can be perceived as the abscissa, i.e. the X-axis of a transversal sinusoidal wave—the light wave. In physics and technology, the light ray model allows comparatively fast and easy description and calculation of the behaviour of light. As shown in the following sections it allows one to characterise the phenomena of refraction, dispersion and reflection. However, intensity-related phenomena or effects that follow from the wave character of light such as diffraction or interference cannot be covered by pure ray optics.

1.2 Refraction

Refraction is the basic underlying mechanism of any classical optical component or system since it describes the change in propagation direction of light waves and rays

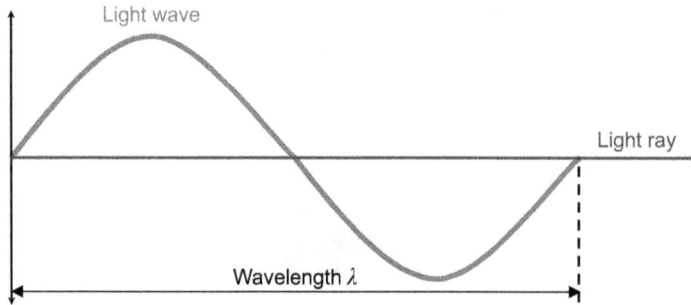

Figure 1.1. Definition of a light ray as the X-axis of a light wave.

at optical interfaces. Here, the essential parameters are the indices of refraction of the involved optical media. The **index of refraction** of a medium n_{medium} is defined by the ratio of the speed of light in vacuum c_0 and the speed of light within the medium c_{medium},

$$n_{\text{medium}} = \frac{c_0}{c_{\text{medium}}}. \tag{1.1}$$

It is thus primarily a measure for the retardation or acceleration of light at the interface of optical media[1]. In terms of the description of optical components and systems, the index of refraction allows one to quantify the change in direction of propagation of light due to the presence of interfaces between different optical media. Refraction of light waves is generally described by the Huygens–Fresnel principle[2] where the wave character of light is taken into account as shown schematically in figure 1.2.

It turns out that according to this principle, the change in propagation direction of a wave front can be described by the normal on the wave front, i.e. a light ray. The refraction of light rays at optical interfaces is generally described by **Snell's law**[3] according to

$$n_1 \cdot \sin \varepsilon_1 = n_2 \cdot \sin \varepsilon_2. \tag{1.2}$$

---

[1] The retardation or acceleration of light at optical interfaces is an amazing effect. For instance, a light wave that comes from ambient air and enters a window pane is deaccelerated from about 300 000 km s^{-1} to approximately 200 000 km s^{-1}, i.e. by a factor of 1.5, corresponding to the index of refraction of window glass. When leaving the glass at the rear side of the window pane, the light wave is instantaneously accelerated to the initial value of 300 000 km s^{-1}.

[2] This principle is named after the Dutch physicist *Christiaan Huygens* (1629–1695) and the French physicist *Augustin-Jean Fresnel* (1788–1827). It states that each point of a wave front is a potential origin of a new fundamental spherical wave.

[3] This basic and important law is named after the Dutch astronomer *Willebrord van Roijen Snell* (1580–1626). However, refraction was first reported by the Greek mathematician *Claudius Ptolemy* (c. 100–170) and described theoretically by the Persian optical engineer *Abu Sad al-Ala ibn Sahl* (c. 940–1000).

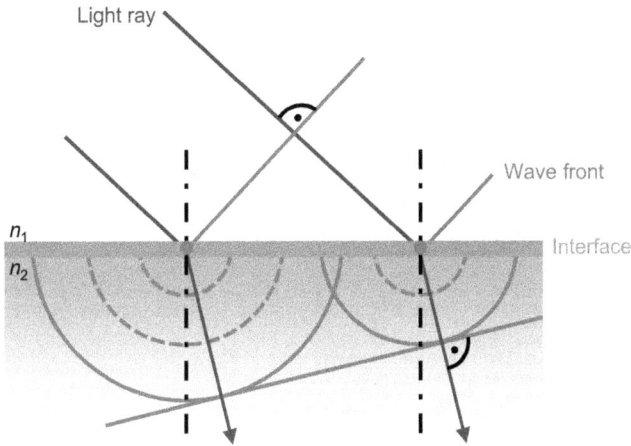

Figure 1.2. Visualisation of refraction according to the Huygens–Fresnel principle.

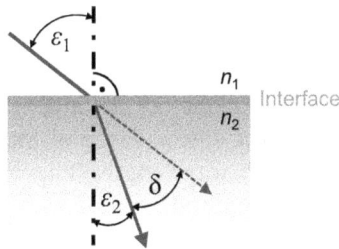

Figure 1.3. Visualisation of Snell's law.

It relates the angle of refraction ε_2 to the angle of incidence of the light ray ε_1 and the index of refraction in front of (n_1) and behind (n_2) the interface as shown in figure 1.3.

The angle of refraction is then given by

$$\varepsilon_2 = \arcsin\left(\frac{n_1 \cdot \sin \varepsilon_1}{n_2}\right) \tag{1.3}$$

and the deviation δ of a light ray from its original direction of propagation follows from

$$\delta = \varepsilon_1 - \varepsilon_2. \tag{1.4}$$

Comprehension question 1.1: Refraction and refractive index

Let us assume an interface between air with $n = 1$ and an optical medium with $n > 1$. A light ray comes from air and crosses the interface. Is it possible to obtain an angle of refraction of 0° if the angle of incidence is higher than 0° (this means that the incident light is inclined and the light ray propagates perpendicularly to the optical interface after refraction)?

Answer: No, this is not possible in practice since the optical medium should feature an infinite index of refraction. This is confirmed by the fact that no solution can be calculated when solving Equation (1.2) for n_2 and inserting the given parameters:

$$n_2 = \frac{n_1 \cdot \sin \varepsilon_1}{\sin \varepsilon_2} = \frac{1 \cdot \sin \varepsilon_1}{\sin 0°} = \frac{\sin \varepsilon_1}{0} = \infty.$$

The actual value of the angle of incidence does not matter as long as it is higher than 0° (= inclined incident light).

1.3 Dispersion

The index of refraction is not a constant value, but features a dependency on wavelength[4]. This wavelength-dependency of the index of refraction is described by the effect of **dispersion**. It is a direct consequence of Snell's law where a particular angle of refraction occurs for each wavelength. This behaviour is the underlying reason for the formation of chromatic aberration (see section 4.2). In the case of normal dispersion, the index of refraction of a transparent medium, as for example glass, decreases with increasing wavelength[5] as shown in figure 1.4. Optical glasses thus feature high indices of refraction for ultraviolet light and lower indices of refraction for visible and (near) infrared light.

For the characterisation of the dispersion characteristics of optical media, several approaches are in hand. For instance, two different parametric descriptions can be applied: First, the **Cauchy equation**[6] and second, the **Sellmeier equation**[7]. Both equations are also known as the dispersion formulas where the first one is given by

$$n(\lambda) = A + \frac{B}{\lambda^2} + ... \tag{1.5}$$

---

[4] The index of refraction of an optical medium does not exclusively depend on wavelength. It is also dependent on the power or intensity of incident incoming light as described by the so-called Kerr coefficient of an optical medium. Further, the index of refraction depends on temperature and pressure as formulated in the famous Edlén equation [4–6], named after the Swedish astrophysicist *Bengt Edlén* (1906–1993). And finally, the index of refraction of anisotropic optical media strongly depends on the polarisation of light.

[5] The effect of increasing index of refraction with increasing wavelength is known as anomalous dispersion.

[6] Named after the French mathematician *Augustin-Louis Cauchy* (1789–1857).

[7] Named after the German physicist *Wolfgang von Sellmeier* (1871).

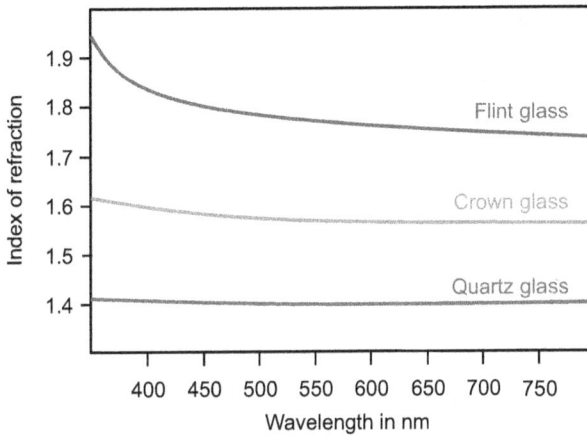

Figure 1.4. Typical dispersion curves of quartz glass, crown glass and flint glass.

Table 1.1. Sellmeier coefficients of selected optical glasses.

	Fused silica	Boron crown glass (Schott, N-BK7)	Heavy flint glass (Schott, SF5)
B_1	0.696 1663	1.039 612 12	1.461 418 85
B_2	0.407 9426	0.231 792 344	0.247 713 019
B_3	0.897 4794	1.010 469 45	0.949 995 832
C_1 (μm^2)	0.068 4043	0.006 000 698 67	0.011 182 612 60
C_2 (μm^2)	0.116 2414	0.020 017 9144	0.050 859 4669
C_3 (μm^2)	9.896 161	103.560 653	112.041 8880
Reference	[1]	[2]	[2]

Here, A and B are the so-called Cauchy parameters, i.e. material-specific coefficients. This equation can be extended by further parameters where at least A and B are required. The index of refraction at any wavelength can thus be calculated if the particular Cauchy parameters of the medium of interest are in hand[8]. This also applies for the Sellmeier equation, given by

$$n(\lambda) = \sqrt{1 + \frac{B_1 \cdot \lambda^2}{\lambda^2 - C_1} + \frac{B_2 \cdot \lambda^2}{\lambda^2 - C_2} + \frac{B_3 \cdot \lambda^2}{\lambda^2 - C_3}}. \tag{1.6}$$

Here, six different coefficients, the Sellmeier coefficients B_1 to C_3 are required. These coefficients are listed for selected optical glasses in table 1.1.

When comparing both approaches, the Cauchy equation and the Sellmeier equation, it turns out that different accuracies can be achieved. Both equations

---

[8] As an example, the Cauchy parameters for the boron crown glass N-BK7 from the glass manufacturer Schott are $A = 1.458\ 00$ and $B = 0.003\ 54$ [7].

allow a quite precise description of the dispersion characteristics of optical media in the visible wavelength range. However, the approach according to Cauchy does not fit very well in the near infrared wavelength range. When choosing one approach, the wavelength or wave band of interest should thus be checked initially.

Even though both approaches are established models for describing the wave-length-dependency of optical media, the dispersion of optical glasses is usually defined and taken into account via the so-called **V-number**[9] in optical design. This value is generally given by the expression

$$V_x = \frac{n_x - 1}{n_y - n_z}. \tag{1.7}$$

It thus refers to a certain wavelength x and follows from the ratio of the index of refraction at this wavelength n_x and the difference of the indices of refraction $n_y - n_z$ at two other wavelengths, y and z. The index of refraction n_x is the index of refraction of the optical medium at the centre wavelength of the considered wavelength band and the difference given in the denominator represents the **main dispersion** of the optical medium. This parameter thus describes the slope of the regression line of a dispersion curve at the two wavelengths y and z.

By default, the standard V-number refers to a centre wavelength of 546.07 nm and the main dispersion given by the indices of refraction at the wavelengths 479.99 and 643.85 nm according to

$$V_e = \frac{n_e - 1}{n_{F'} - n_{C'}}. \tag{1.8}$$

As listed in table 1.2, these wavelengths correspond to the Fraunhofer lines[10], which are commonly marked by the indices e, F' and C'. One has to consider that this standard was defined some years ago[11]. Before, the V-number was given on the basis of the index of refraction at the Fraunhofer line d and the main dispersion at the Fraunhofer lines F and C according to

$$V_d = \frac{n_d - 1}{n_F - n_C}. \tag{1.9}$$

The V-number allows a fast evaluation of the dispersion characteristics of optical glasses and is an important parameter for the calculation of achromatic systems

---

[9] Sometimes (and especially in German-speaking countries), the V-number is also referred to as the Abbe-number, named after the German physicist and optical scientist *Ernst Karl Abbe* (1840–1905), a pioneer in the theoretical description of optical glasses and technical optics.

[10] In 1814, the German optician and physicist *Joseph von Fraunhofer* (1787–1826) reported on dark lines in the solar spectrum. These lines were first observed by the English chemist and physicist *William Hyde Wollaston* (1766–1828) in 1802 and result from self-absorption of sunlight within the plasma atmosphere of the Sun. Here, each element has a specific and discrete absorption wavelength.

[11] The V-number was re-defined since the Fraunhofer line e ($\approx$546 nm), which is now used as the standard centre wavelength, is better adapted to the maximum sensitivity of the human eye at daylight, i.e. 555 nm.

Table 1.2. Fraunhofer lines and the corresponding wavelengths used for the calculation of the V-numbers V_e and V_d.

Fraunhofer line	Wavelength (nm)
e	546.07
F'	479.99
C'	643.85
d	587.56
F	486.13
C	656.27

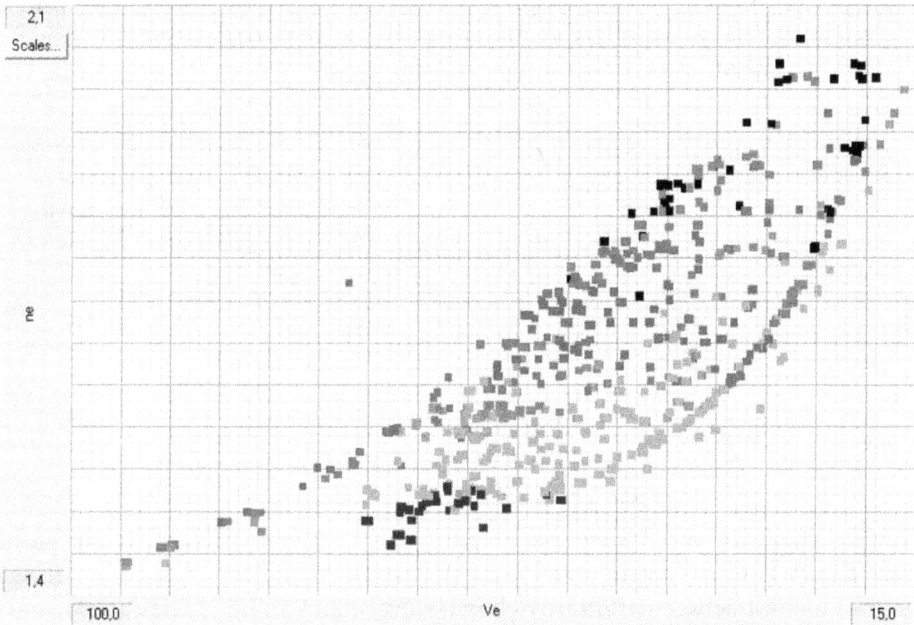

Figure 1.5. n–V-diagram, a.k.a. Abbe diagram, as provided in the software WinLens.

(see section 3.6). Optical glass catalogues thus normally contain a so-called **n–V-diagram**, a.k.a. **Abbe diagram**, shown in figure 1.5.

Here, the index of refraction at the centre wavelength e or d is plotted versus the V-number at the same wavelength. One has to note that the V-numbers are given in descending order on the X-axis of this diagram since the lower the V-number of an optical medium, the higher its dispersion. This diagram allows the categorisation of glasses into the two main glass families, **crown glass** and **flint glass**. The separation between both glass families is found at a V-number of $V_e = 50$[12] and the distribution is as follows:

---

[12] For an n–V-diagram plotting n_d vs V_d, the separation is found at $V_d = 55$.

- $V_e > 50 \rightarrow$ crown glass with a low index of refraction, a high V-number and thus a low dispersion
- $V_e < 50 \rightarrow$ flint glass with a high index of refraction, a low V-number and thus a high dispersion[13].

Usually, converging lenses are made of crown glass whereas flint glass is used for diverging lenses.

Comprehension question 1.2: Refraction index and V−number

Figure 1.5 shows an n–V-diagram where the index of refraction is plotted versus the V-number. The displayed point cloud represents the available glasses. Which type of glass would be appreciated for optical system design and where would it be found in the n—V-diagram?

Answer: A glass with a high index of refraction, but low dispersion and thus a high V-number is desirable and would be helpful for optics design. It would be found top left in the n-V-diagram. But as we can see, there is no such glass existent due to the basic underlying interrelationship: The higher the index of refraction, the lower the V-number.

FOCUS ON:

Using the n–V- diagram

$\rightarrow$ Exercise **E27**

Another value for the description of the dispersion characteristics is the **partial dispersion** P. It allows one to describe the dispersion characteristics for a wavelength range of interest between arbitrarily-chosen wavelengths x and y, respectively, and is given by

$$P_x = \frac{n_x - n_y}{n_{F'} - n_{C'}} \tag{1.10}$$

or

$$P_x = \frac{n_x - n_y}{n_F - n_C}. \tag{1.11}$$

Again, the denominator represents the particular main dispersion. Partial dispersion characterises the slope of the regression line between the two chosen wavelengths within the dispersion curve of the optical material of interest. It is thus an indicator for the bending of the dispersion curve.

---

[13] Flint glasses with a very low V-number and high dispersion are referred to as heavy or dense flint glasses.

1.4 Reflection

Reflection is an important effect that should be considered in lens design or optical system design. On the one hand, reflection at glass surfaces leads to a reduction in transmitted light as described in more detail in section 1.5. It therefore directly affects the luminosity of an imaging system. On the other hand, reflected light can super-impose and interfere, giving rise to the formation of ghost images (see section 4.4) or high local light intensity within the optical system[14].

The amount of reflected light or reflectance R depends on a number of parameters: the indices of refraction n_1 and n_2 of the involved optical media, the angle of incidence ε_1 and the angle of refraction ε_2 as well as the direction of polarisation of light[15]. This interrelationship is specified by the so-called **Fresnel equations**. For perpendicular polarised light the reflectance R_s[16] is given by

$$R_s = \left(\frac{n_1 \cdot \cos \varepsilon_1 - n_2 \cdot \cos \varepsilon_2}{n_1 \cdot \cos \varepsilon_1 + n_2 \cdot \cos \varepsilon_2} \right)^2. \tag{1.12}$$

For parallel polarised light, the reflectance R_p follows from

$$R_P = \left(\frac{n_2 \cdot \cos \varepsilon_1 - n_1 \cdot \cos \varepsilon_2}{n_2 \cdot \cos \varepsilon_1 + n_1 \cdot \cos \varepsilon_2} \right)^2. \tag{1.13}$$

A special case occurs for normal incidence of light where where $\varepsilon_1 = 0$. Here, a simplified approximation can be applied for the calculation of the reflectance R:

$$R = \left(\frac{n_2 - n_1}{n_2 + n_1} \right)^2. \tag{1.14}$$

As an estimation, this approximation is valid for small angles of incidence lower than approximately 10°.

Reflection features two special properties that are relevant in optical system design. First, merely the perpendicular polarised fraction of the non-polarised incoming light is reflected at a specific angle of incidence, the so-called **Brewster angle**[17]. It is given by

$$\varepsilon_{\text{Brewster}} = \arctan\left(\frac{n_2}{n_1} \right) \tag{1.15}$$

---

[14] High local light intensity due to the interference of laser irradiation may lead to laser-induced damage of an optical system, e.g. laser objectives.

[15] The direction of polarisation of light gives the oscillation direction of the field vector of the electric field of a light wave.

[16] Note that the index 's' originates from the German word 'senkrecht' (= 'perpendicular' in English). This index is still given by the initial letter of the German term since the English terms both start with a 'p' ('perpendicular' and 'parallel').

[17] The Brewster angle is named after the Scottish physicist *Sir David Brewster* (1781–1868) who described the impact of surface reflection on the polarisation of light in 1815 [8].

with n_1 and n_2 being the index of refraction in front of the interface and behind it, respectively. This effect is technically used for realising simple polarisers with high damage threshold as, for example, required in laser cavities. Second, a further special phenomenon occurs at the interface from an optically thick to an optically thinner medium, e.g. from glass to air. Here, incident light is totally reflected[18] if its angle of incidence corresponds to or exceeds the so-called **critical angle** of total internal reflection, given by

$$\varepsilon_{\text{crit}} = \arcsin\left(\frac{n_1}{n_2}\right). \tag{1.16}$$

Note that here n_1 is the index of refraction behind the surface and n_2 is the index of refraction in front of the surface. The effect of total internal reflection has significant importance and is applied in a number of different optical components and systems such as step index fibres or—referring to imaging optical systems—deflecting prisms such as Porro prism pairs[19], which are found in modern binoculars.

1.5 Absorption and transmission

In order to achieve a high brightness of the image, as much light as possible should be transmitted by the optical components of an optical imaging system. Apart from surface reflection, the **absorption** A or the internal **transmission** T_i, respectively, of optical media is thus of great interest. When passing through an optical medium, the intensity of light is subject to an exponential decay. This behaviour is described by the **Beer–Lambert law**[20]. It relates the intensity of light I_x at any position x within an optical medium to the initial intensity of light[21] I_0 according to

$$I_x = I_0 \cdot e^{-\alpha \cdot x}. \tag{1.17}$$

Here, α is the **absorption coefficient**[22] of the optical medium. The expression

$$e^{-\alpha \cdot x} = \frac{I_x}{I_0} = T_i \tag{1.18}$$

is the internal transmission T_i. This parameter allows one to calculate the absorbance A of an optical medium according to

---

[18] Actually, there is no real total reflection at the infinite thin optical interface. Since light is an electromagnetic wave, its electric field overshoots into the optical medium behind the interface where the depth of penetration is in the order of magnitude of wavelength, depending on the angle of incidence. The irradiated sector within the neighbouring medium is referred to as the evanescent field.

[19] Named after the Italian engineer *Ignazio Porro* (1801–1875).

[20] The Beer–Lambert law is named after the Swiss mathematician *Johann Heinrich Lambert* (1728–1777) and the German mathematician *August Beer* (1825–1863) who published this law in 1760 [9] and 1852 [10], respectively. However, it is based on older work reported by the French physicist *Pierre Bouguer* (1698–1758) in 1729 [11].

[21] Note that the initial intensity of light I_0 is given by the intensity of light after entering the optical medium, so it is the residual intensity after reflection at the interface.

[22] The reciprocal value of the absorption coefficient is—by definition—the optical penetration depth ($d_{\text{opt}} = 1/\alpha$). At this depth, the intensity of incoming light is attenuated by the factor $1/e \approx 0.37$.

$$A = 1 - T_i. \tag{1.19}$$

For the determination of the intensity of light transmitted by an optical component, both surface reflection and internal absorption have to be taken into account. The total transmission T_{tot} thus accounts for

$$T_{tot} = (1 - R)^2 \cdot e^{-a \cdot x} \tag{1.20}$$

if the reflectance R at the front surface of the component corresponds to the reflectance R at its rear side[23].

One has to consider that according to the Fresnel equations, surface reflectance is strongly polarisation-dependent. This dependency thus also applies for the transmission T of light at an optical interface, which can be calculated according to

$$T_s = \left(\frac{2 \cdot n_1 \cdot \cos \varepsilon_1}{n_1 \cdot \cos \varepsilon_1 + n_2 \cdot \cos \varepsilon_2} \right)^2 \tag{1.21}$$

for perpendicular polarised light and

$$T_P = \left(\frac{2 \cdot n_1 \cdot \cos \varepsilon_1}{n_2 \cdot \cos \varepsilon_1 + n_1 \cdot \cos \varepsilon_2} \right)^2 \tag{1.22}$$

for parallel polarised light. It should finally be mentioned that in addition to surface reflection and intrinsic absorption within the (theoretically perfect) bulk material, further losses in light intensity may occur due to diffuse surface reflection caused by surface roughness[24] and due to absorption or scattering at impurities and bubbles within an optical medium. However, these effects are usually not considered in the course of basic optical design.

1.6 Interference

As mentioned in section 1.2, light can be described as a transversal electromagnetic wave. It is thus subject to specific wave phenomena as, for example, **interference**. Interference results from the superposition of waves where two characteristic cases can occur: constructive and destructive interference. The first case describes the superposition of, for example, two peaks or wave crests in phase. Here, the particular amplitudes of both crests accumulate—the resulting wave is amplified.

In contrast, destructive interference occurs in the case of superposition of a wave crest and a wave trough. The amplitude of the resulting wave is then attenuated and even annihilated for perfect constructive interference. This physical principle is the

---

[23] This is valid for a constant surrounding medium, normally air. However, equation (1.20) does not apply for optical components where the index of refraction in front of the component differs from the index of refraction behind the component, since here each interface features its particular difference in the index of refraction and, as a consequence, reflectance. As an example, this applies for the front lens of underwater cameras or the front lens of a microscope objective that is in contact with immersion liquid in the object space.

[24] The impact of surface roughness on the refection behaviour is taken into account by the so-called total integrated scatter (TIS). This function gives the amount of diffusively scattered light [12].

basis of dielectric coatings[25] used in optical systems. Against this background, both the index of refraction n_l and the thickness t_l of the applied layer are of prime importance. For normal incidence, the thickness of an antireflective coating is given by

$$t_l = \frac{\lambda}{4 \cdot n_l}. \tag{1.23}$$

For angles of incidence $0° < \varepsilon < 90°$, it follows from

$$t_l = \frac{\lambda}{4 \cdot \sqrt{n_l^2 - \sin^2 \varepsilon}}. \tag{1.24}$$

A dielectric mirror coating is obtained when adapting the prefactor in the denominator. Equations (1.23) and (1.24) are then rewritten; the layer thickness at normal incidence is

$$t_l = \frac{\lambda}{2 \cdot n_l} \tag{1.25}$$

and

$$t_l = \frac{\lambda}{2 \cdot \sqrt{n_l^2 - \sin^2 \varepsilon}} \tag{1.26}$$

respectively, for other angles of incidence[26].

Case example 1.1: Comparison of blank and coated optics

Task
The total transmittance of an optical system consisting of six lenses shall amount to at least 80%. As an estimation, we assume that the index of refraction of all the involved lenses is 1.5 and that the lenses are not cemented. Further, internal absorption within the lenses is disregarded.

Solution
The total transmittance T_{tot} of an optical system is given by

$$T_{tot} = T_{surf}^N$$

where T_{surf} is the transmittance of a single surface and N is the number of surfaces[27]. The required total transmittance is $80\% = 0.8$. The acceptable transmittance per surface is thus

---

[25] In contrast to metallic coatings, dielectric coatings are transparent layers or stacks of several layers where magnesium fluoride (MgF_2) and silicon dioxide (SiO_2) are established and common optical materials.
[26] It should be noted that equations (1.23)–(1.26) are only valid if the index of refraction of the surrounding medium is 1.
[27] This equation is valid if all involved surfaces feature the same transmittance.

$$T_{\text{surf}} = \sqrt[12]{T_{\text{tot}}} = \sqrt[12]{0.8} = 0.9816 = 98.16\%.$$

The required total transmittance can thus be achieved by applying an antireflective coating with a maximum residual reflectance of 1.84% to each surface.

For comparison, the total transmittance of the optical system without any coating is

$$T_{\text{tot}} = T_{\text{surf}}^{12} = 0.96^{12} = 0.61 = 61\%$$

when assuming normal incidence (AOI = 0°) at all surfaces since in this case,

$$T_{\text{surf}} = 1 - R_{\text{surf}} = 1 - \left(\frac{n_2 - n_1}{n_2 + n_1}\right)^2 = 1 - 0.04 = 0.96.$$

1.7 Diffraction and optical resolution

Another phenomenon that occurs due to the wave character of light is **diffraction**. Broadly speaking, this effect describes the propagation of light into shadowed sectors behind barriers. According to the model of light rays, these sectors could not be reached but in practice, propagation of light occurs as a consequence of the Huygens–Fresnel principle (compare section 1.3) since a new spherical wave is formed at the point of contact of the incident wave front and the edge of the barrier as shown in figure 1.6.

The superposition of the new spherical waves formed gives rise to constructive and destructive interference, finally leading to the formation of a diffraction pattern. The shape of this pattern depends on the geometry of the aperture that was passed by the incident light. In In figure 1.6, a rotational-symmetric diffraction pattern that follows from a circular aperture is shown as an example. This interference pattern is

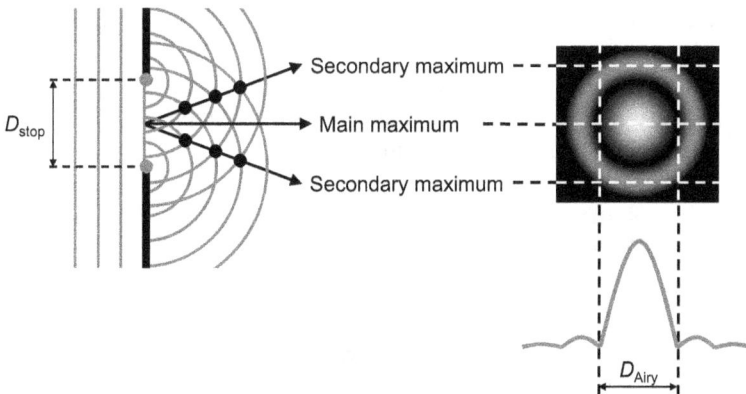

Figure 1.6. Visualisation of the formation of an Airy disc due to diffraction and interference including the definition of the Airy disc diameter D_{Airy}.

commonly known as the so-called **Airy disc**[28]. It has significant importance for the design and evaluation of classical imaging optical systems: An object point that is imaged on a detector is thus not given by an infinite small point as suggested by ray optics, but an Airy disc with intensity maxima and minima. The diameter D_{Airy} of this pattern is given by the diameter of the first intensity minimum and can be calculated according to

$$D_{\text{Airy}} = 1.22 \cdot \frac{\lambda \cdot f}{D_{\text{stop}}}. \tag{1.27}$$

Here, λ is the wavelength of light, f is the focal length of the imaging optical system, and D_{stop} is its stop diameter[29]. This diameter is of essential importance since it gives the physical limit of resolution. Optical systems that theoretically produce image points with diameters smaller than the Airy disc diameter are referred to as diffraction-limited systems. Moreover, two image points can be clearly distinguished if the distance between the corresponding discs is equal to or higher than half the Airy disc diameter[30] as shown in figure 1.7.

> **FOCUS ON:**
> Airy disc diameter
> → Exercise **E9** and **E34**

This rule is known as the **Rayleigh criterion**[31] [3]. It is fulfilled if the first intensity minimum of one Airy disc coincides with the principal maximum of the other one. The consideration of diffraction-related phenomena is thus of essential importance for the design of imaging optical systems.

1.8 Fundamentals—bottom line

- The propagation of light can be described by a wave or a ray.
- Refraction at optical interfaces, the basis of optical imaging, is described by Snell's law.
- According to the Huygens–Fresnel principle each point of a wave front is a potential origin of a new fundamental spherical wave.
- The index of refraction of a medium is defined by the ratio of the speed of light in vacuum and the speed of light within the medium.
- Dispersion is the effect of splitting incoming white light into its spectral components by an optical medium.

---

[28] Named after the English mathematician and astronomer *Sir George Biddell Airy* (1801–1892).
[29] Note that equation (1.27) is only valid if the distance from the optical system to the detector is in the order of the focal length of the optical system.
[30] Strictly speaking, this definition only applies to monochromatic light since the Airy disc diameter depends on wavelength.
[31] Named after the British scientist *John William Strutt*, 3rd Baron Rayleigh (1842–1919).

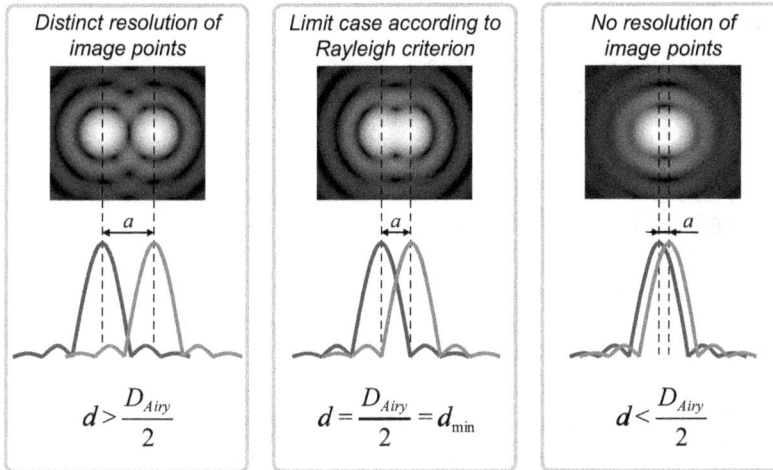

Distinct resolution of image points	Limit case according to Rayleigh criterion	No resolution of image points
a	a	a
$d > \dfrac{D_{Airy}}{2}$	$d = \dfrac{D_{Airy}}{2} = d_{min}$	$d < \dfrac{D_{Airy}}{2}$

Figure 1.7. Visualisation of the Rayleigh criterion.

- A medium's dispersion characteristics can be described by the Cauchy equation, the Sellmeier equation, the V-number or partial dispersion.
- The current standard wavelength or centre wavelength of a wave band for describing optical media and components is 546.07 nm (Fraunhofer line e).
- The classification of optical glasses is performed by the n–V-diagram.
- Crown glasses feature a low index of refraction, a high V-number and thus a low dispersion.
- Flint glasses feature a high index of refraction, a low V-number and thus a high dispersion.
- Reflection is described by the Fresnel equations where the reflectance follows from the indices of refraction of the involved optical media at an interface, the angle of incidence and the polarisation of light.
- Absorption within optical media is described by the Beer–Lambert law.
- Interference is the amplification or extinction of light due to the superposition of light waves.
- Dielectric optical layers such as antireflective coatings are based on interference.
- Diffraction occurs due to the wave character of light and describes the formation of new fundamental waves at barriers such as stops according to the Huygens–Fresnel principle.
- Diffraction leads to the formation of an intensity pattern, the so-called Airy disc.
- Optical systems producing image points smaller than the Airy disc diameter are diffraction-limited.
- According to the Rayleigh criterion, two image points can be resolved spatially if the distance between the points is at least half the Airy disc diameter.

References

[1] Maltison I 1965 *J. Opt. Soc. Am.* **55** 1205–8

[2] Schott A G 2014 Optisches Glas—Datenblätter (in German)

[3] Lord Rayleigh F R S 1879 *Philos. Mag.* **5** 261–27

[4] Edlén B 1966 *Metrologia* **2** 71–80

[5] Birch K P and Downs M J 1993 *Metrologia* **30** 155–62

[6] Ciddor P E 1996 *Appl. Opt.* **35** 1566–73

[7] Schott A G 2015 Optical Glass Data Sheets

[8] Brewster D 1815 *Philos. Trans. R. Soc. Lond.* **105** 125–59

[9] Lambert J H 1760 *Photometria, sive de mensura et gradibus luminis, colorum et umbrae* 1st edn (Augsburg: Sumptibus Vidae Eberhardi Klett)

[10] Beer A 1852 *Ann. Phys. Chem.* **86** 78–88 (in German)

[11] Bouguer P 1729 *Essai d'optique, sur la gradation de la lumière* 1st edn (Paris: Claude Jombert)

[12] Harvey J E, Choi N, Schroeder S and Duparré A 2012 *Opt. Eng.* **51** 013402

IOP Publishing

Lens Design Basics
Optical design problem-solving in theory and practice
Christoph Gerhard

Chapter 2

Introduction to imaging models

To this day, the character of light is not yet fully understood. Usually, wave-particle dualism is applied for the description of light. According to this approach, light is an electromagnetic wave as characterised by a sinusoidal wave with its proper wavelength and amplitude on the one hand. It is thus subject to wave phenomena such as diffraction or interference. On the other hand, it consists of particles—photons—with a certain particle energy. And according to the German physician *Max Planck* (1858–1947), light is neither a wave nor a particle but exhibits particular characteristics. In optical system design, a much simpler approach is used for classical imaging tasks: ray optics. Here, two cases are differentiated, the paraxial model for light rays close to the optical axis and the geometric-optical imaging model for large fields of view. Both models as well as the basics of wave optics are presented in this chapter.

2.1 The paraxial or Gaussian imaging model

An interesting fact is that most children intuitively paint the Sun with sunrays, even though the Sun does not emit sunrays. However, the description of light by light rays thus seems to be a subliminal approach. This approach is applied in **paraxial imaging**. This imaging model is an extremely simplified and thus idealised description of the behaviour of light propagating close to the optical axis of an imaging system [1]. This approach considers merely reflection and refraction at optical interfaces whereas phenomena related to the wave character of light, i.e. diffraction and interference, are neglected. The light paths are described by light rays[1]; paraxial imaging is thus sometimes also referred to as **ray optics**. An essential feature of this imaging model is the restriction of the field of view [2], expressed by the aperture angle u or the field angle ω, as introduced in figure 2.1. As long as these angles do not

---

[1] A light ray is the normal on a wave front.

doi:10.1088/978-0-7503-2240-9ch2

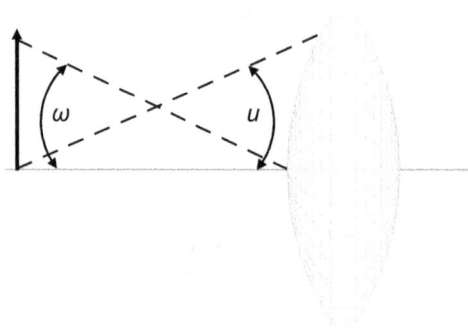

Figure 2.1. Definition of the aperture angle u or the field angle ω.

exceed a value of 5°, the paraxial imaging model can be applied[2]. Since the limited imaging space covered by an angle of ⩽5° is referred to as the **Gaussian image space**, it is thus also known as the **Gaussian imaging** model[3,4].

This model stands out due to its easy mathematical handling since object points and image points are definitely correlated. Hence, it allows a quick overview on the relationships and interdependencies in optical systems where optical imaging can be described by so-called cardinal elements as, for example, the focal points, the principal points and the nodal points as introduced in more detail in section 3.1.3. Moreover, further reference or auxiliary parameters, e.g. the focal length or the magnification, can be defined. Such parameters are helpful values for the description of imaging tasks at higher aperture or field angles.

The application of the paraxial or Gaussian imaging model further gives an essential reference plane in optics manufacturing: the **Gaussian image plane**. For a given imaging task where the object distance and height as well as the focal length of the imaging system are given, the image distance as calculated via the paraxial imaging model defines the position of this plane where—theoretically—the image is formed[5].

2.2 Geometric-optical imaging

The **geometric-optical imaging** model considers 'widely-opened' systems, i.e. optical systems with a large field of view (FOV) and high aperture and field angles, respectively[6]. It thus addresses the sector where the aperture angle or field angle are

---

[2] In this case, the small-angle approximation can be applied, so $u = \sin u = \tan u$ and $\omega = \sin \omega = \tan \omega$.

[3] Named after the German mathematician *Johann Carl Friedrich Gauß* (1777–1855).

[4] Not to be confused with Gaussian beam propagation, i.e. a wave optical description of the propagation of laser beams.

[5] Strictly speaking, the image distance and the position of the Gaussian image plane, respectively, as calculated via the paraxial imaging model, is only valid for thin lenses, i.e. lenses that are described by a single principal plane (see section 3.1.2). In practice and for real lenses, the real or true image plane is shifted from the Gaussian image plane. This effect is referred to as 'defocus'. Finally, the image plane is usually not a plane, but a curved surface.

[6] For instance, the binocular field of vision of an adult or the 'optical system human eyes', respectively, is approximately 214° in a horizontal direction and 130°–150° in a vertical direction.

Object at infinity
→
Aperture angle = 0°

Increase in defocus

Increase in defocus

Scale 1.0mm   Spacing 1.0mm   Def. 0.0mm
-2.0mm   -1.0mm   .0mm   1.0mm   2.0mm
on-axis

Finite conjugates
→
Aperture angle > 5°

Scale 1.0mm   Spacing 1.0mm   Def. 0.0mm
-2.0mm   -1.0mm   .0mm   1.0mm   2.0mm
on-axis

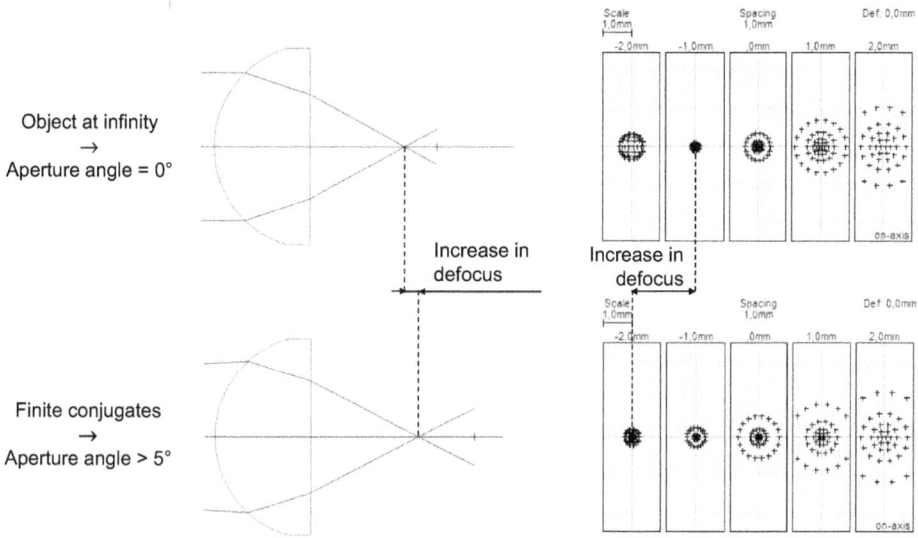

Figure 2.2. Visualisation of the impact of aperture angle on the position of the focal point and the smallest image spot, respectively.

higher than 5°. This sector is also known as the Seidel space[7]. Since the nominal focal length of a lens is defined for light coming from infinity—and thus featuring an angle of 0° with respect to the optical axis—a certain shift in focal point, the so-called defocus[8] occurs for light coming from this sector as visualised in figure 2.2. As a consequence, the smallest image point is not found at the Gaussian image plane, but shifted towards the lens[9].

Moreover, larger aperture and field angles and consequently inclined incidence of light leads to the appearance of field-related aberrations such as coma and distortion as presented in more detail in chapter 4 and shown in figures 2.3 and 2.4[10]. In this example, it can be seen that the spot diagram shows both coma and astigmatism and the transverse ray aberration diagram further indicates distortion for the full field (here: 30°). In addition, the transmitted wave front distortion is significantly higher for the full field as shown by the comparison in figure 2.4.

The mathematical handling or calculation of optical systems using this approach is relatively easy and the geometric interrelations described by this model represent the basis for the construction of optical systems. It is thus sufficient for a considerable number of classical optical setups. The most important parameters used for calculation

---

[7] Named after the German mathematician and optician *Philipp Ludwig von Seidel* (1821–1896).

[8] A certain defocus also occurs for light propagating parallel to the optical axis in the object space. The larger the aperture, the higher the defocus since the angle of refraction and thus the aperture angle in the image space —a.k.a. the image angle—increases.

[9] In some cases, the focal point position can also be shifted towards the image plane.

[10] The diagrams shown in the figures, i.e. the spot diagram, the transverse ray aberration diagram and the 3D wave front plot are introduced in chapter 5.

Spot diagram

Transverse ray aberration diagram

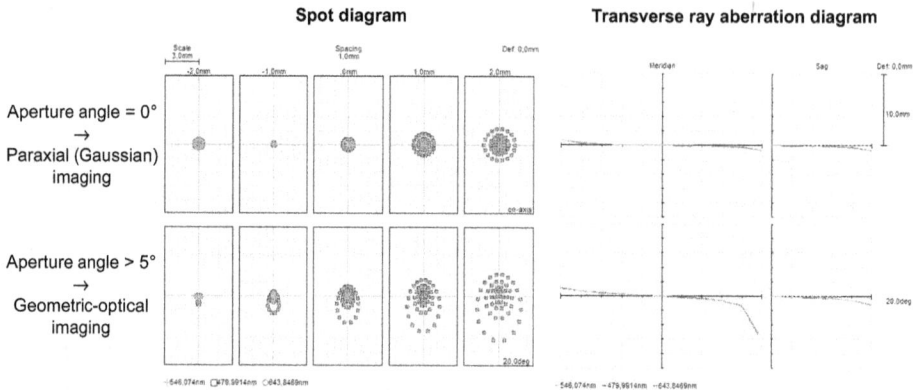

Aperture angle = 0°
→
Paraxial (Gaussian)
imaging

Aperture angle > 5°
→
Geometric-optical
imaging

Figure 2.3. Visualisation of the formation of field-related aberrations via the spot diagram and the transverse ray aberration diagram.

Aperture angle = 0°
→
Paraxial (Gaussian) imaging

Aperture angle > 5°
→
Geometric-optical imaging

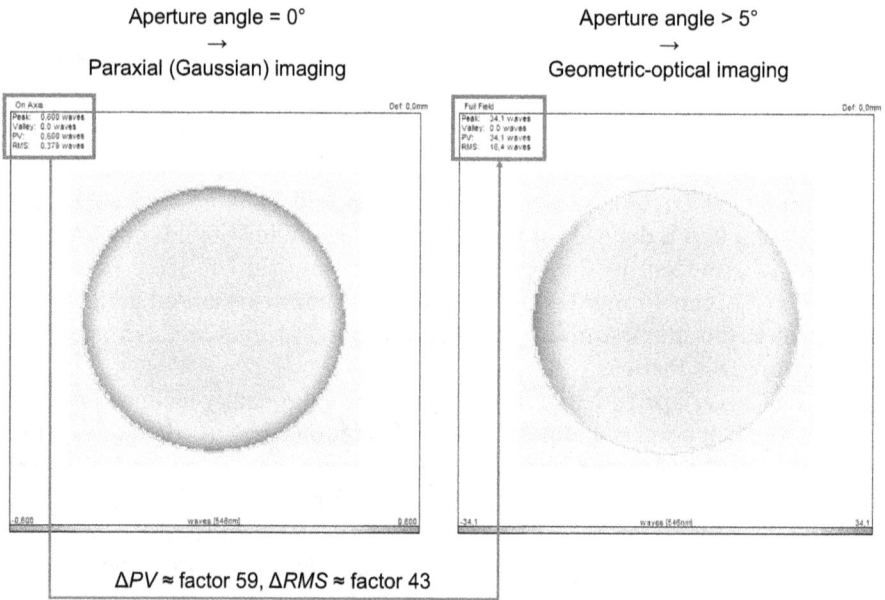

$\Delta PV \approx$ factor 59, $\Delta RMS \approx$ factor 43

Figure 2.4. Comparison of the transmitted wave front for on-axis light and the full field including the differences in peak-to-valley value (PV) and root mean square (rms) deviation.

are the ray entrance height and the aperture angle as described in more detail in section 4.1.1.

> FOCUS ON:
> Paraxial and geometric-optical imaging
> → Exercise **E6**

2.3 Wave optics

Both the paraxial and the geometric-optical imaging model neglect the wave character of light. This essential character is considered by **wave optics**, a.k.a. **physical optics**; it allows one to clarify the limits of optical resolution that occur even in the case of perfect paraxial or geometric-optical imaging. Moreover, other wave optical phenomena such as diffraction at stops and diaphragms, the formation of diffraction patterns in the focal point of—primary—coherent light as well as interferences within the optical path due to reflections and backscattering at lens and mount surfaces can be described. As shown in figure 2.5, wave optics is also required for the depiction of the functional principle of dielectric coatings.

Wave optics is also suitable for the characterisation and calculation of non-classical optical systems as, for example, holograms [3, 4] or diffractive optical elements (DOE) [5]. Finally, the propagation of laser beams through an optical system is described based on wave optics. Even though small aperture angles apply for theoretically collimated laser beams, the paraxial imaging model is not applicable in this case. This fact is visualised in figure 2.6.

It can be seen that for the paraxial or geometric-optical imaging model, an infinite small focus diameter is found in the focal point. In reality, a focused laser beam—or any focused light—features a certain focus diameter that results from the wave character of light. As described in more detail in section 1.7, the lower limit of the focus diameter is given by the Airy disc diameter.

FOCUS ON:
Ray optics vs. wave optics
→ Exercise **E11**

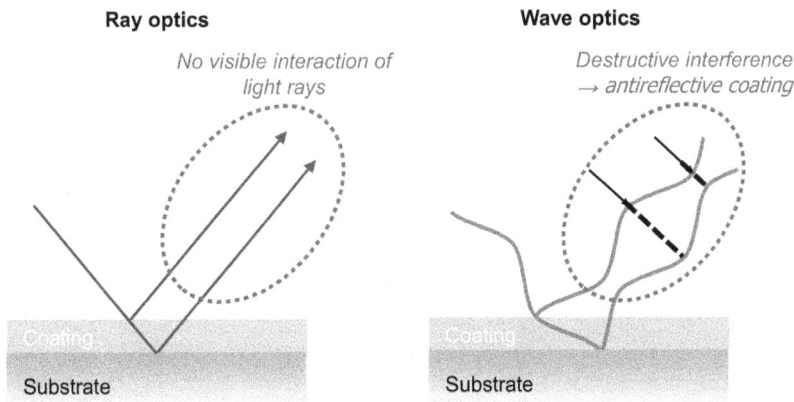

Ray optics

No visible interaction of light rays

Wave optics

Destructive interference → antireflective coating

Coating

Substrate

Coating

Substrate

Figure 2.5. Comparison of the ray optical and the wave optical model; for the description of interference effects at dielectric coatings, the latter model is required.

Ray optics

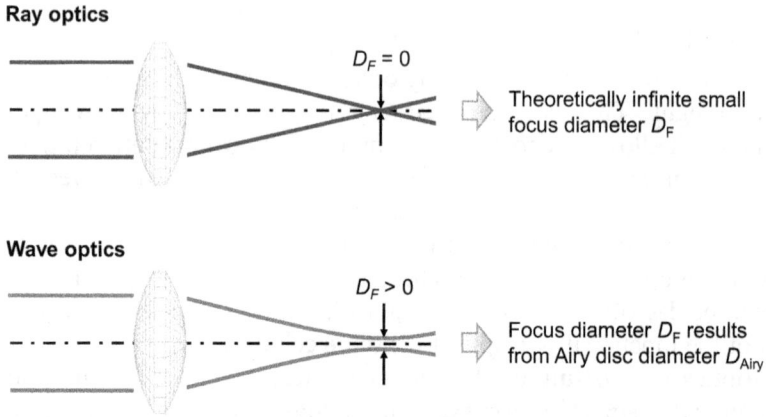

$D_F = 0$

Theoretically infinite small
focus diameter D_F

Wave optics

$D_F > 0$

Focus diameter D_F results
from Airy disc diameter D_{Airy}

Figure 2.6. Comparison of the ray optical and the wave optical model for the determination of focus or image point diameters; an infinite small diameter as determined by the ray optical model is impossible in practice.

Case example 2.1: Defocus due to wave character of light

Task

A laser beam with a wavelength of $\lambda = 633$ nm and a beam waist diameter of $2w_0 = 8$ mm is focused by a converging lens with a focal length of $f = 100$ mm. The beam waist of the raw beam is located ten metres in front of the lens. Our goal is now to determine the position of the focal point after focusing.

Solution

As a first approach, the position of the focal point amounts to the nominal focal length of 100 mm according to the paraxial imaging model. Since the object distance is 10 m, the incoming light can be considered as coming from infinity[11]. In reality, the focal point position differs from this nominal value due to the wave character of light and the fact that for the calculation of laser light paths, diffraction has to be taken into account. Here, the approach of Gaussian beam propagation is applied. According to this approach, the difference in focal length and actual focal point position, respectively, Δf follows from

$$\Delta f = \left(\frac{d \cdot f^2}{d^2 + z_R^2} \right) \tag{2.1}$$

where $d = 10$ m and z_R is the Rayleigh length, given by

$$z_R = \frac{\pi \cdot w_0^2}{\lambda}. \tag{2.2}$$

---

[11] As an approach, object distances >2 m can be considered to correspond to infinity for first estimations using the paraxial imaging model.

It amounts to

$$z_R = \frac{\pi \cdot (4 \text{ mm})^2}{0.000\ 633 \text{ mm}} \approx 79.4 \text{ m}.$$

The shift in focal length and focal point position is thus

$$\Delta f = \left(\frac{10 \text{ m} \cdot (0.1 \text{ mm})^2}{(10 \text{ m})^2 + (79.4 \text{ mm})^2} \right) \approx 16 \text{ μm}.$$

The focal point is thus shifted by approximately 16 μm. One has to consider that this comparatively low value is only valid for a perfect laser beam with a beam quality number of $M^2 = 1$.

2.4 Introduction to imaging models—bottom line

- Paraxial imaging represents the basic imaging model but is only valid for small aperture and field angles ($<5°$).
- Paraxial imaging considers refraction and reflection close to the optical axis.
- Geometric-optical imaging is valid for large fields of view (aperture and field angles $>5°$).
- Geometric-optical imaging considers refraction and reflection for extensive apertures.
- Wave optics takes interference and diffraction phenomena into account and is thus suitable for describing optical coatings and the resolution of imaging.

References

[1] Lipson A, Lipson S G and Lipson H 2010 *Optical Physics* 4th edn (Cambridge: Cambridge University Press)
[2] Greivenkamp J E 2004 *Field Guide to Geometrical Optics* 1st edn (Bellingham, WA: SPIE press)
[3] Gabor D 1948 *Nature* **161** 777–8
[4] Leith E and Upatnieks J 1964 *J. Opt. Soc. Am.* **54** 1295–301
[5] Testorf M 2006 *Appl. Opt.* **45** 76–82

IOP Publishing

Lens Design Basics
Optical design problem-solving in theory and practice
Christoph Gerhard

Chapter 3

Calculation of simple optical components and systems

The calculation of complex optical systems is a challenging and extensive task. In recent times it was performed manually and when comparing an old daguerreotype[1] to a modern digital or analogue image, it turns out that the image quality has not been notably enhanced in the last 170 years. However, the calculation of an optical system is nowadays mainly carried out with the aid of simulation tools that reduce the required expenditure of work and time significantly. Both approaches, the manual and the computer-assisted methods, are based on the same physical interrelations and equations. This chapter gives an overview of the basic mathematical models for the calculation of simple optical components such as single lenses and systems, e.g. achromatic doublets. Such components represent the basis of any optical systems.

3.1 Lens types and basic lens parameters

3.1.1 Lens shapes

Single lenses can be categorised in quite different ways:
- based on the lens functionality, i.e. converging or diverging,
- based on the used material, i.e. glass lenses, crystal lenses, liquid lenses etc, or
- based on the shape of the optically active surface.

According to the latter categorisation, cylindrical lenses, toric lenses, conical lenses[2], aspherical lenses and spherical lenses can be defined where the latter type is

---

[1] Daguerreotype: the first photograph or photographic process that was suitable for mass production, named after its inventor, the French photographer *Louis-Jacques-Mandé Daguerre* (1787–1851).
[2] Conical lenses are usually also referred to as axicons.

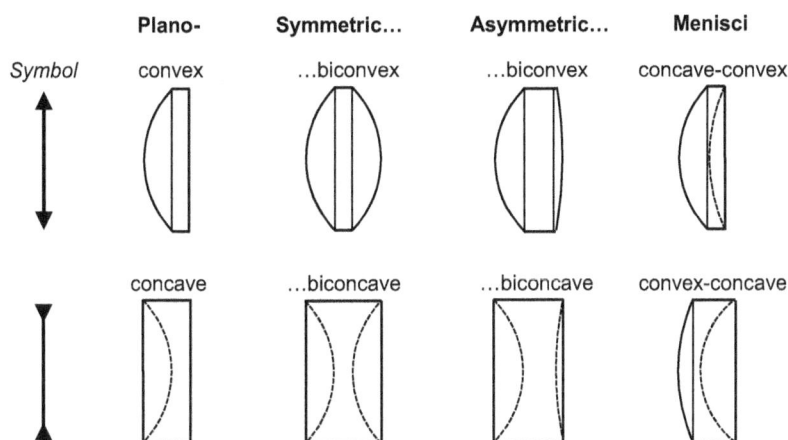

	Plano-	Symmetric...	Asymmetric...	Menisci
Symbol	convex	...biconvex	...biconvex	concave-convex
	concave	...biconcave	...biconcave	convex-concave

Figure 3.1. Different types of spherical lenses—converging lenses (top) and diverging lenses (bottom)—including the particular symbol.

the classical one[3]. Here, at least one surface is given by a ball segment. Depending on the direction of curvature, different types of spherical lenses can be defined as shown in figure 3.1.

It can be seen that the orientation of the particular radius of curvature gives the general shape and thus type of lens. This orientation is thus of specific importance and interest for the design and manufacture of lenses. It is expressed by the algebraic sign of the radius of curvature as listed in table 3.1.

Apart from the radii of curvature, further parameters are needed for an entire definition of an optical lens as visualised in figure 3.2: the centre thickness t_c, the lens diameter D, and the lens material, i.e. the glass or other optical medium with its proper index of refraction n, absorption coefficient α and V-number V.

> **FOCUS ON:**
> Definition of optical lenses
> → Exercise **E12**

For basic considerations in optical system design, the actual shape of a lens might not be of importance. Fur such considerations and simple calculations, the main function of the lens is sufficient. This means that the focal length including its algebraic sign have to be known in order to define the optical power[4] as well as the behaviour or function of a lens: converging lenses feature a positive algebraic sign

---

[3] Spherical lenses are quite easy to manufacture (with respect to the other lens shapes) and have been produced for millennia; the oldest presently known lens is dated around 1000 BCE. It is called the Nimrud lens, named after the place where it was found, the ancient Assyrian city Nimrud in northern Mesopotamia.

[4] The optical power P is given by the the reciprocal of the focal length ($P = 1/f$).

Table 3.1. Different types of spherical lenses including sign definition of the particular radius of curvature (if arranged towards incoming light as shown in figure 3.1) and respective characteristic.

Lens type	Sign definition		Characteristic
	First (left) surface	Second (right) surface	
Plano-convex	+	(∞)	$R_2 = \infty$
Symmetric biconvex	+	—	$R_1 = R_2$
Asymmetric biconvex (best form lens)	+	—	$R_1 \neq R_2$
Positive (concave-convex) meniscus	+	+	R_1 (convex) $< R_2$ (concave)
Plano-concave	+	(∞)	$R_2 = \infty$
Symmetric biconcave	+	—	$R_1 = R_2$
Asymmetric biconcave	+	—	$R_1 \neq R_2$
Negative (convex-concave) meniscus	+	+	R_1 (convex) $> R_2$ (concave)

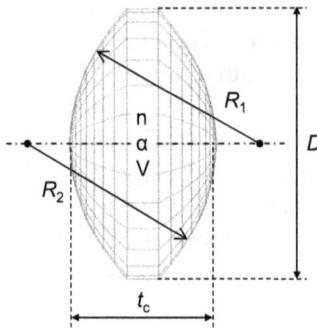

Figure 3.2. Lens parameters required for the definition of a lens: radii of curvature R, centre thickness t_c, diameter D and lens material, specified by the index of refraction n, the absorption coefficient α and the V-number V.

whereas the focal length of diverging lenses is negative. The basic behaviour of lenses is visualised by the symbols shown in figure 3.1.

3.1.2 Principal planes

Any imaging optical component or system can be described by two virtual planes, the so-called **principal planes**, where refraction occurs theoretically. The position of such principal planes can be constructed graphically on the basis of a simple rule: after passing through a physical lens, the light ray that initially propagates parallel to the optical axis crosses the focal point and vice versa (for the description of construction rays see section 3.3.2). This transformation is due to refraction at the lens surfaces. When extrapolating the paths of all rays in front and behind a lens or optical system, virtual intersection points occur as shown in figure 3.3.

The planes that are given by the connection of these points are referred to as the principal planes that are orientated perpendicularly to the optical axis. Two principal planes can be identified, one on the object side (H) and one on the image side (H'). Depending on the shape of a physical lens or system, the principal planes can even be located outside the lens material as shown by some examples in figure 3.4.

The principal planes are of essential importance since the effective focal length of a lens refers to this plane and not to the surface of a lens. For an idealised and simplified description of any imaging system, even merely one principal plane is sufficient for the estimation of the relationships between the object space and image space of an optical system[5].

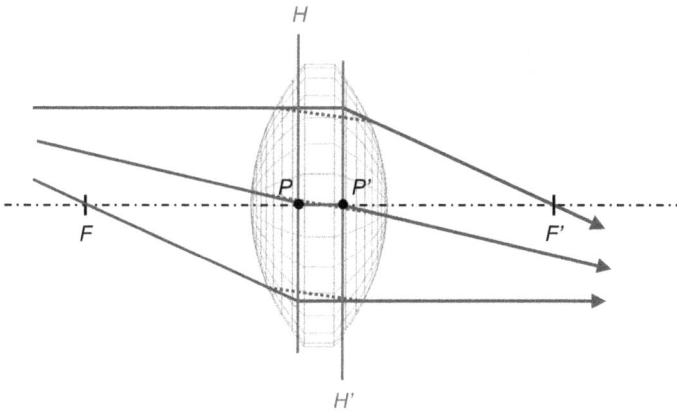

Figure 3.3. Graphical construction of the principal planes H and H' of a biconvex lens.

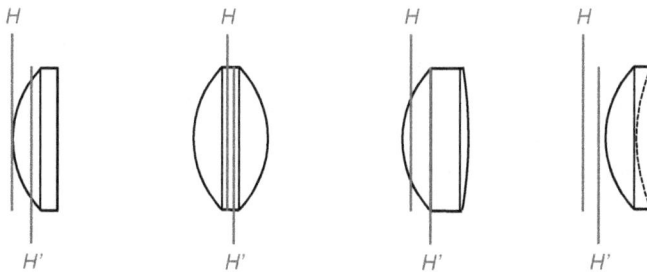

Figure 3.4. Qualitative visualisation of the location of principal planes for different converging lens types.

---

[5] This approach—reducing an optical system to merely one principal plane—is applied in the PreDesigner software used for solving the exercises and examples in this book.

Table 3.2. Overview on cardinal elements including the particular description.

Element	Abbreviation	Description
Principal points	P, P'	Intercept points of principal planes and optical axis (in case of axially-parallel incident light)
Vertices	V, V'	Intercept points of lens surfaces and optical axis
Focal points	F, F'	Front and rear focal points
Effective focal length	EFL	Focal length referring to the principal plane (commonly called focal length f)
Back focal length	BFL	Focal length referring to vertex in the image space
Front focal length	FFL	Focal length referring to vertex in the object space
Nodal points	N, N'	Equivalent to principal points in case of inclined incidence of light

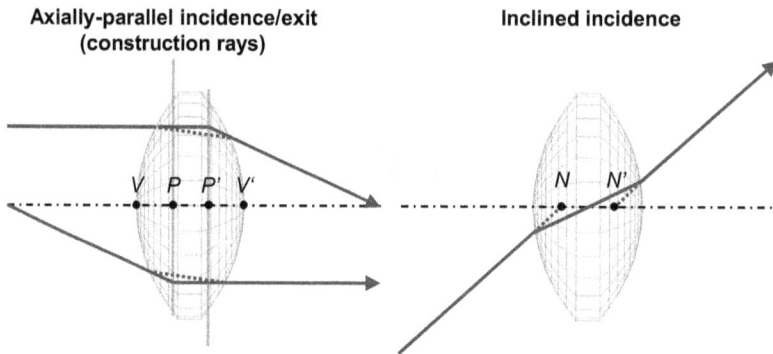

Figure 3.5. Cardinal elements of a biconvex lens as listed in table 3.2 for axially-parallel light and inclined light.

3.1.3 Cardinal elements

Apart from the above-mentioned principal planes, further characteristic positions or points on lens surfaces or within the lens material as well as distances can be identified. These points and distances are the so-called *cardinal elements* as listed in table 3.2 and partially shown in figure 3.5.

Cardinal elements are of essential importance since the calculation and layout design of any imaging system is based on these parameters. For instance, the radii of curvature of a physical lens follows from the focal length as described in the following section.

3.2 The lensmaker's equations

The **lensmaker's equations** allow the determination of the required parameters for realising a physical lens with a certain focal length (and vice versa). As summarised in figure 3.2, the essential parameters of a lens are the index of refraction (and absorption coefficient and V-number) of the lens material, the radii of curvature and the centre thickness. It turns out that the calculation of these parameters is based on several assumptions due to the comparatively high number of variables. Here, professional experience and intuition become of importance and sometimes, simple economic considerations may already define some of the desired parameters[6].

3.2.1 Thin lenses

In many cases, it is sufficient to start optical design with a very simple approach: the **thin lens**. Even though such a lens does not really exist in practice, it is a helpful model for first estimations and calculations of an optical system. By definition, the centre thickness of a thin lens is much smaller than the radii of curvature, $t_c \ll R_1, R_2$. As a consequence, the lens can be described as a single principal plane since the principal planes of both surfaces are congruent due to the marginal centre thickness. The effective focal length of such a thin lens generally follows from

$$\frac{1}{\mathrm{EFL}} = (n_l - 1) \cdot \left(\frac{1}{R_1} - \frac{1}{R_2} \right). \tag{3.1}$$

Here, n_l is the index of refraction of the lens material[7]. For equi-curved symmetric biconvex or biconcave lenses, equation (3.1) simplifies to

$$\frac{1}{\mathrm{EFL}} = \frac{2 \cdot (n_l - 1)}{R} \tag{3.2}$$

and even

$$\frac{1}{\mathrm{EFL}} = \frac{(n_l - 1)}{R} \tag{3.3}$$

for plano-convex or plano-concave thin lenses.

3.2.2 Thick lenses

In real life, the thickness of a lens cannot be neglected. This especially applies for lenses where the centre thickness is in the order of magnitude of the radii of curvature as in, for example, hemispherical lenses. Such lenses are referred to as **thick lenses**. The principal planes of both optical interfaces are thus separated and

---

[6] For instance, the choice of the radii of curvature may be driven by the existing and available standard tools for optics manufacturing. Further, the glass and the index of refraction, respectively, may be chosen on the basis of availability and pricing.
[7] Please note that equation (3.1) and the other lensmaker's equations presented here do not consider the index of refraction of the ambient medium (which is thus defined to amount to 1 by default).

the centre thickness t_c has to be considered during calculation. Equation (3.1) is then rewritten as follows:

$$\frac{1}{\text{EFL}} = (n_l - 1) \cdot \left(\frac{1}{R_1} - \frac{1}{R_2} + \frac{(n_l - 1) \cdot t_c}{n_l \cdot R_1 \cdot R_2} \right). \tag{3.4}$$

> FOCUS ON:
> Lensmaker's equations
> → Exercises **E13** and **E14**

3.3 Conjugated parameters of optical imaging

3.3.1 Definition of conjugated parameters

There are two different relevant sectors in optical imaging. The **object space** in front of a lens or system and the **image space** behind it. The essential parameters in the object space and the corresponding ones in the image space are so-called **conjugated parameters** that are interconnected via the focal length or the magnification, respectively, of the imaging optics (see section 3.4). As shown in figure 3.6, imaging thus means the transfer of an object with a certain height y and a certain distance to the optical system a to an image with its proper image height y' and distance a' that follow from the characteristics of the used optics[8]. These conjugated parameters and other essential factors that are relevant for imaging are listed in table 3.3.

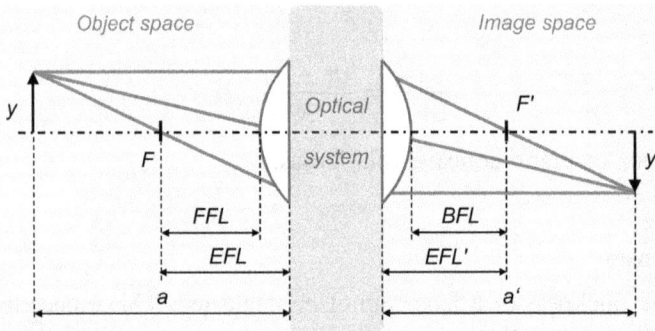

Figure 3.6. Schematic of optical imaging by an optical system or component including the relevant, partially conjugated parameters as listed in table 3.3.

---

[8] Note that in German-speaking countries, the object height and distance are abbreviated by the letters 'G' and 'g', respectively, originating from the word 'Gegenstand' ('object' in English). Moreover, the formula symbols of the image height and distance are 'B' and 'b', respectively, since the German word for 'image' is 'Bild'. These abbreviations are often found in the literature.

Table 3.3. Overview of the essential parameters for optical imaging. The particular heights and distances are conjugated parameters.

Object space		Image space	
Abbreviation	Parameter	Abbreviation	Parameter
y (or u)	Object height	y' (or u')	Image height
a	Object distance	a'	Image distance
EFL	Effective front focal length	EFL$'$	Effective back focal length
FFL	Front focal length	BFL	Back focal length
F	Focal point	F'	Focal point
H	Principal plane	H'	Principal plane

One has to consider that the effective focal lengths refer to the particular principal plane whereas the back focal length and the front focal length refer to the front and rear vertices of the lens[9].

3.3.2 Construction rays

For the graphical construction of optical imaging through a lens or system with a given focal length at least two so-called **construction rays** are required. There are three different construction rays. First, the **parallel ray** propagates parallel to the optical axis in the object space. After passing the lens, it crosses the focal point in the image space (→ 'the parallel ray becomes the focal ray'). Second, the **focal ray** crosses the focal point in the object space and propagates parallel to the optical axis after passing the lens (→ 'the focal ray becomes the parallel ray'). Third and finally, the **chief ray** passes the lens without any alteration of its direction. It is thus not subject to changes in propagation angle with respect to the optical axis. However, a certain longitudinal offset, depending on the centre thickness of the lens occurs. An overview of the three construction rays is given in figure 3.7.

Both the parallel ray and the focal ray were already mentioned in the course of the definition of principal planes in section 3.1.2. Construction rays thus allow for the graphical construction of imaging tasks and the identification of essential positions within optical systems.

3.4 The imaging equation and magnification

The general **imaging equation** describes the interrelationship of the effective focal length EFL and the object distance a and the image distance a'[10] according to the reciprocal addition

---

[9] As a consequence, the object and image distances always refer to the principal planes of a lens. For calculating the physical air gap between the object and the lens' front surface or the image plane and the lens' rear surface, the sagitta (a.k.a. camber) of the lens has thus to be taken into account. This point is of interest for the assembly of opto-mechanical systems.

[10] Note that for a real imaging, the image height is negative if the object height is positive and vice versa.

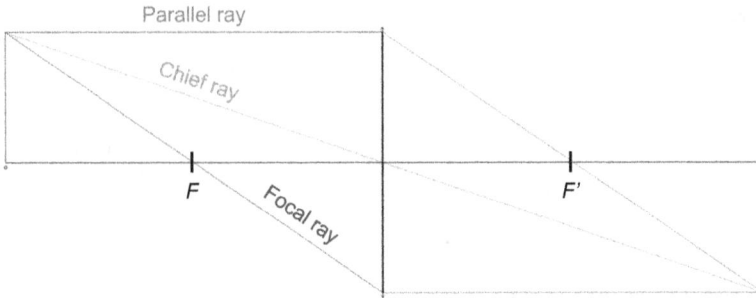

Figure 3.7. Definition of the construction rays—the parallel ray, the focal ray and the chief ray—as displayed in the software tool PreDesigner. Note that here the lens or optical system is represented by a single principal plane.

$$\frac{1}{\text{EFL}} = \frac{1}{a} - \frac{1}{a'}. \tag{3.5}$$

The object and image distance are thus conjugated or connected by the focal length. This means that for given heights and distances in the object space, a specific image height and distance is resulting and vice versa.

FOCUS ON:	FOCUS ON:
Conjugated parameters	Object and image height
→ Exercise **E1**	→ Exercise **E2**

Conjugated parameters can also be linked by the so-called **Newton's equation** according to

$$\text{EFL} \cdot \text{EFL}' = z \cdot z' = (a - \text{EFL}) \cdot (a' - \text{EFL}'). \tag{3.6}$$

Another essential parameter which does not only link the object distance a and the image distance a', but also the object height y and the image height y', is the **magnification** m of an optical system[11]. It is given by

$$m = \frac{a'}{a} = \frac{y'}{y}. \tag{3.7}$$

In practice, this parameter is usually given and turns out to be very helpful, for example, for the determination of the object distance if the image distance is unknown:

$$a = \text{EFL} \cdot \left(1 - \frac{1}{m}\right) \tag{3.8}$$

---

[11] Note that in German-speaking countries, the formula symbol of the magnification m is usually the Greek letter beta (β).

or vice versa:

$$a' = -\text{EFL} \cdot (1 - m). \tag{3.9}$$

3.5 Calculation of compound lenses

Up to now, we have considered the most basic optical element, a single lens. However, optical systems consist of at least two, but usually more lenses. A classical and simple optical element is a **compound lens** made of two single lenses. Such a setup can be realised by air-gapped lenses or cemented[12] ones, so-called **doublets**. In the latter case, no air gap is thus existent between both lenses. The total effective focal length EFL_{tot} of a lens doublet follows from

$$\frac{1}{\text{EFL}_{\text{tot}}} = \frac{1}{\text{EFL}_1} + \frac{1}{\text{EFL}_2} - \frac{d}{\text{EFL}_1 \cdot \text{EFL}_2} \tag{3.10}$$

or

$$\text{EFL}_{\text{tot}} = \frac{\text{EFL}_1 \cdot \text{EFL}_2}{\text{EFL} + \text{EFL}_2 - d}, \tag{3.11}$$

respectively. Here, EFL_1 is the effective focal length of the first lens, EFL_2 the effective focal length of the second one and d is the distance between both lenses[13]. The total magnification m_{tot} of such lens doublets is the product of the single magnifications of the involved lenses and is thus given by

$$m_{\text{tot}} = m_1 \cdot m_2. \tag{3.12}$$

It can also be expressed via the object and image distances according to

$$m_{\text{tot}} = \frac{a'_1}{a_1} \cdot \frac{a'_2}{a_2} = \frac{a'_1 \cdot a'_2}{a_1 \cdot (d - a'_1)}. \tag{3.13}$$

For more complex systems, i.e. triplets or optical systems consisting of four and more lenses, complex ray tracing becomes necessary as described extensively in [1–4]. This task is nowadays solved via computer-based design as presented in more detail in chapter 6.

3.6 Condition for achromatism

A special type of lens doublet is the so-called **achromatic doublet** or **achromatic lens**[14]. This simple optical system addresses the following effect: as a consequence of dispersion (see section 1.3), white light passing through a single lens is split into its spectral fractions. This phenomenon is referred to as chromatic aberration as

---

[12] In practice, lenses are connected by so-called optical fine cements in some cases, e.g. achromatic doublets. Such cements are normally two-compound polymer liquids that are cured and hardened via UV-irradiation.
[13] Strictly speaking, d is the distance between the principal planes of both lenses.
[14] The achromatic lens was invented by the English optician John Dollond (1706–1761) in 1758.

presented in more detail in section 4.2. For two selected wavelengths, dispersion can quite easily be compensated by the combination of two lenses with different V-numbers and thus different dispersion characteristics. As shown in section 3.5, the focal length of such an achromatic doublet follows from the focal lengths of the involved single lenses, a converging one and a diverging one. Compensation of two-wavelength chromatic aberration occurs if the so-called **condition for achromatism** is fulfilled. According to this general condition, the absolute values of the product of the effective focal length EFL and the V-number V of the first involved lens has to correspond to the product of the effective focal length and the V-number of the second one:

$$\text{EFL}_1 \cdot V_1 = -\text{EFL}_2 \cdot V_2. \tag{3.14}$$

The algebraic signs of this general condition show that an achromatic lens consists of a converging lens and a diverging one. Both lenses are usually cemented, so the distance is $d = 0$. The total effective focal length of the achromatic doublet EFL_{al} is thus given by

$$\frac{1}{\text{EFL}_{\text{al}}} = \frac{1}{\text{EFL}_1} + \frac{1}{\text{EFL}_2}. \tag{3.15}$$

FOCUS ON:
Condition for achromatism
$\rightarrow$ Exercise **E22**

For a default or predetermined effective focal length of an achromatic doublet, the particular parameters of the involved single lenses can be determined quite easily: solving the general condition for achromatism given by equation (3.14) for the effective focal length of the second lens gives

$$\text{EFL}'2 = \frac{\text{EFL}'1 \cdot V_1}{V_2}. \tag{3.16}$$

Inserting this expression in equation (3.15) results in

$$\frac{1}{\text{EFL}_{\text{al}}} = \frac{1}{\text{EFL}_1} \cdot \frac{V_1 - V_2}{V_1}. \tag{3.17}$$

This expression can then be solved for the effective focal length of the first lens,

$$\text{EFL}_1 = \text{EFL}_{\text{al}} \cdot \frac{V_1 - V_2}{V_1}, \tag{3.18}$$

or the second lens,

$$\text{EFL}_2 = -\text{EFL}_{\text{al}} \cdot \frac{V_1 - V_2}{V_2}. \tag{3.19}$$

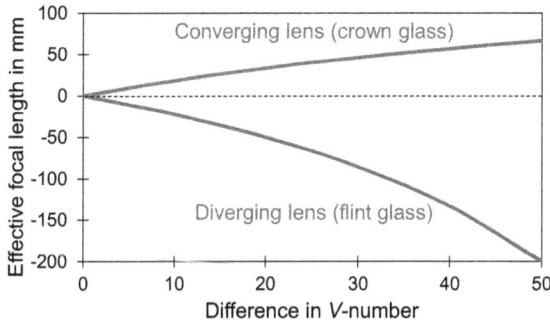

Figure 3.8. Visualisation of the increase in effective focal length with increasing difference in V-number for an achromatic doublet with a total focal length of 100 mm.

Once the focal lengths of both lenses are determined, the geometric lens parameters can be calculated on the basis of the lensmaker's equations (see section 3.2). The indices of refraction required here are indirectly defined by the V-numbers and the type of glass used for both lenses. The choice of the glasses is an interesting and challenging task where some experience in optics design is useful. Normally, a standard crown glass[15] with a high V-number and a low dispersion, respectively, is used as an approach for the converging lens of an achromatic doublet. For the diverging lens, a flint or heavy flint glass with a low V-number and a high dispersion, respectively, is chosen. As a simple rule, the difference in V-number of both glasses should be as high as possible since the effective focal lengths of the involved single lenses increase with increasing difference in V-number as shown in figure 3.8.

As a consequence of the lensmaker's equations, the radii of curvature increase with increasing effective focal length. This finally leads to a reduction of a basic optical image defect, spherical aberration (see section 4.1.2), due to a decrease in particular angles of incidence and refraction at the lens surfaces. The use of an achromatic doublet thus not only allows for a correction of chromatic aberration, but also spherical aberration to some degree.

Case example 3.1: Calculation of an achromatic lens

Task
A simple optical system with a total focal length of 100 mm should be corrected for two wavelengths, 479.99 and 643.85 nm. We thus have to calculate a possible solution for such a system.

Solution
A simple system for solving the given task is an achromatic doublet. The given wavelengths correspond to the Fraunhofer lines *F'* and *C'*. Our consideration is thus

---

[15] Standard glasses have several advantages: such glasses are competitive, readily available and quite easy to machine. These points are of interest since pricing, manufacturing, practicability and sustainability are important secondary issues in optical system design!

Table 3.4. Optical properties of the optical glasses N-BK7 and N-SF6. Data taken from [5].

	Property	Glass
	N-BK7® (crown glass)	N-SF6 (flint glass)
n_e	1.518 72	1.812 66
n_F	1.522 83	1.829 80
$n_{C'}$	1.514 72	1.797 49
V_e	63.96	25.16

performed based on the V-number defined by the main dispersion n_F-$n_{C'}$ according to equation (1.8). The doublet does not necessarily have to be a cemented lens group. We thus use a simple approach: the combination of a thin symmetric biconvex lens made of low-dispersive crown glass and a thin plano-concave one consisting of a high-dispersive flint or heavy flint glass. In order to obtain lenses with high radii of curvature and low spherical aberration we choose a high difference in V-number as visualised in figure 3.8. Moreover, we should choose standard glasses on account of availability, machinability and cost. Such glasses are, for example, N-BK7® and N-SF6, both from Schott. The optical parameters of these glasses are listed in table 3.4. We see that the difference in V-number is quite high, i.e. $\Delta V_e = 38.8$.

The actual calculation can now be peformed in several successive steps:

Step 1: Determination of particular effective focal lengths
According to equation (3.18), we obtain the required effective focal length of the first lens, i.e. the thin symmetric biconvex one:

$$EFL_1 = 100 \text{ mm} \cdot \frac{(63.96 - 25.16)}{63.96} = +60.66 \text{ mm}.$$

The effective focal length of the second lens follows from equation (3.19):

$$EFL_2 = (-100 \text{ mm}) \cdot \frac{(63.96 - 25.16)}{25.16} = -154.21 \text{ mm}.$$

Since this lens is a thin plano-concave one, its focal length is negative.

Step 2: Determination of particular radii of curvature
The absolute value of the radii of curvature of the symmetric biconvex lens can be determined by solving equation (3.2) for R (here: R_1 of the first lens):

$$\frac{1}{EFL_1} = \frac{2 \cdot (n_l - 1)}{R_1} \rightarrow R_1 = 2 \cdot (n_l - 1) \cdot EFL_1$$
$$= 2 \cdot (1.518 72 - 1) \cdot 60.66 \text{ mm} = 62.39 \text{ mm}.$$

Here, we insert n_e for the index of refraction of the lens material n_l. The calculated radius of curvature thus applies to the centre wavelength of the wave band[16] defined by V-number V_e. The radius of curvature of the plano-convex lens follows from equation (3.3):

$$\frac{1}{EFL_2} = \frac{(n_l - 1)}{R_2} \rightarrow R_2 = (n_l - 1) \cdot EFL_2$$
$$= (1.812\,66 - 1) \cdot (-154.21\ \text{mm}) = -125.32\ \text{mm}.$$

Step 3: Determination of wavelength-dependent total focal lengths
According to equation (3.15) and thus assuming no air gap between the lenses, we obtain the target total effective focal length of 100 mm at the centre wavelength of the wave band, 546.07 nm:

$$\frac{1}{EFL_{al}} = \frac{1}{60.66\ \text{mm}} + \frac{1}{(-154.21\ \text{mm})} \rightarrow EFL_{al} = 100\ \text{mm}.$$

For 479.99 nm, i.e. Fraunhofer line F', we obtain the following particular focal lengths:

$$EFL_1 = \frac{R_1}{2 \cdot \left(n_{F'(\text{crown glass})} - 1\right)} = \frac{62.39\ \text{mm}}{2 \cdot (1.522\,83 - 1)} = 59.67\ \text{mm}$$

and

$$EFL_2 = \frac{R_2}{n_{F'(\text{flint glass})} - 1} = \frac{-125.32\ \text{mm}}{1.829\,80 - 1} = -151.02\ \text{mm}.$$

The total focal length of the achromatic doublet at a wavelength of 479.99 nm is then

$$\frac{1}{EFL_{al}(F')} = \frac{1}{59.67\ \text{mm}} + \frac{1}{(-151.02\ \text{mm})} \rightarrow EFL_{al}(F') = 98.67\ \text{mm}.$$

For Fraunhofer line C' or a wavelength of 643.85 nm, we obtain

$$EFL_1 = \frac{R_1}{2 \cdot \left(n_{C'(\text{crown glass})} - 1\right)} = \frac{62.39\ \text{mm}}{2 \cdot (1.514\,72 - 1)} = 60.61\ \text{mm},$$

$$EFL_2 = \frac{R_2}{n_{C'(\text{flint glass})} - 1} = \frac{-125.32\ \text{mm}}{1.797\,49 - 1} = -157.14\ \text{mm},$$

and

$$\frac{1}{EFL_{al}(C')} = \frac{1}{60.61\ \text{mm}} + \frac{1}{(-157.14\ \text{mm})} \rightarrow EFL_{al}(C') = 98.67\ \text{mm}.$$

---

[16] Not only the radii of curvature, but also the defined effective focal length always apply to the centre wavelength of the considered wave band.

We see that we get the same effective focal length of the doublet for both wavelengths of interest, 479.99 and 643.85 nm. This doublet is thus a possible solution for the correction of chromatic aberration for those wavelengths.

> **Comprehension question 3.1: Achromatic doublet**
>
> The effective focal length of the achromatic doublet in Case example 3.1 is 98.67 mm at 479.99 nm and 643.85 nm. However, at the centre wavelength of 546.07 nm, it amounts to the nominal value, 100 mm. What is the reason for this difference of 1.33 mm, and what is this difference called in optics design?
>
> *see section 4.2.*
> *the used glasses. This residual chromatic aberration is called secondary spectrum,*
> *wavelength, including the centre wavelength, due to the dispersion characteristics of*
> *residual chromatic aberration for the corrected wavelengths and any other*
> *Answer: Even though the doublet is corrected for two wavelengths, it still features*

3.7 Calculation of aplanatic lenses

Aplanatic lenses are a special type of optical system consisting of at least two single lenses. Such an aplanatic system is free of spherical aberration and coma. In practice, aplanatic menisci are often used as accessory lenses for increasing the field of view of existing optical systems. Such an aplanatic meniscus images a small field close to the optical axis free of aberrations. Its first or front surface images even a complete sphere free of aberrations and its second or back surface is concentric around the on-axis image point.

Case example 3.2: Calculation of an aplanatic meniscus

Task

The numerical aperture of a microscope lens amounts to $NA = 0.10$. It should be increased by an accessory aplanatic meniscus with the given parameters:
- index of refraction: 1.65,
- centre thickness: 1 mm,
- distance from principal plane of the microscope lens to the substrate plane: 29 mm,
- working distance: 25 mm.

We are thus looking for the radii of curvature of the aplanatic meniscus.

Solution

The second or back surface of an aplanatic meniscus is free of any aberrations if its radius of curvature R_2 corresponds to its distance to the image point d_2. It is then concentric around the image point, the Seidel coefficients (see section 4.1.1) consequently amount to zero. Moreover, the same condition must be fulfilled at the first surface. This is true or valid if the distance from the second vertex of the meniscus to the image point d_1 amounts to

$$d_1 = \frac{n_2 + n_1}{c_1 \cdot n_1} \tag{3.20}$$

with n_1 being the index of refraction of the medium in front of the fist surface, i.e. air with $n_1 = 1$, n_2 being the index of refraction of the lens material and the coefficient c_1 being the reciprocal value of the radius of curvature $(1/R_1)$. The distance is quite easy to determine, it amounts to

$$d_1 = 29 \text{ mm} - 4 \text{ mm} = 25 \text{ mm}.$$

The value of 4 mm is the difference of the distance from the principal plane of the microscope lens to the substrate plane (29 mm) and the working distance (25 mm). Solving equation (3.20) for c_1 gives

$$c_1 = \frac{n_2 + n_1}{d_1 \cdot n_1} = \frac{1.65 + 1}{25 \text{ mm} \cdot 1} = 0.106 \text{ mm}^{-1}.$$

The radius of curvature of the first surface thus amounts to $1/c_1 = 9.4$ mm. The radius of curvature of the second lens surface R_2 follows from

$$R_2 = d_2 = \left(\frac{u_1 \cdot d_1}{u_2} \right) - t_c. \tag{3.21}$$

The distance from the second vertex of the meniscus to the image point d_1 was just calculated, and the centre thickness t_c and the aperture angle u_1 are given, the latter by the actual numerical aperture NA_1. The new numerical aperture of the microscope lens after adding the accessory aplanatic meniscus NA_2 can be determined as follows:

$$n_2 \cdot NA_2 = n_1 \cdot NA_1 + d_1 \cdot NA_1 \cdot (n_2 - n_1) \cdot c_1$$

$$\rightarrow \quad NA_2 = \frac{0.272\,25}{1.65} = 0.165.$$

The radius of curvature of the second lens surface R_2 and the new working distance d_2, respectively, are then

$$R_2 = d_2 = \left(\frac{NA_1 \cdot d_1}{NA_2} \right) - t_c = \left(\frac{0.1 \cdot 25 \text{ mm}}{0.165} \right) - 1 \text{ mm} = 14.15 \text{ mm}.$$

By adding such an accessory aplanatic meniscus, the working distance is thus reduced by a factor of 1.77 and the numerical aperture or aperture angle was increased by a factor of 1.65.

3.8 Calculation of simple optical systems—bottom line

- In terms of shape, spherical lenses are the standard lenses used for optical imaging.
- For a complete definition of a spherical lens, its radii of curvature, centre thickness, diameter and material are required.
- Any optical component or system can be described by two virtual planes, the so-called principal planes.
- The effective focal length of a lens refers to the particular principal plane.
- A lens or optical system features a number of characteristic points, the so-called cardinal elements.
- Cardinal elements are the principal points, the vertices, the focal points, the effective focal length, the back and front focal length as well as the nodal points.
- The lensmaker's equations allow for the determination of the focal length based on geometrical lens parameters and the index of refraction of the lens material.
- By definition, the centre thickness of a thin lens is much smaller than the radii of curvature.
- For thick lenses, the centre thickness has to be taken into account.
- The essential parameters for optical imaging are the object height and distance as well as the image height and distance.
- The essential parameters in the object space and the corresponding ones in the image space are so-called conjugated parameters.
- Conjugated parameters are interconnected via the focal length or the magnification, respectively.
- The graphical construction of optical imaging through a lens or system can be performed with the aid of construction rays.
- There are three different construction rays,
  - the parallel ray,
  - the focal ray, and
  - the chief ray.
- During imaging,
  - the parallel ray becomes the focal ray,
  - the focal ray becomes the parallel ray, and
  - the chief does not change its direction.
- The imaging equation describes the interrelationship of the effective focal length and the object and image distance.
- The magnification is given by the ratio of the image distance and the object distance or the ratio of the image height and the object height.
- The calculation of achromatic doublets is based on the condition for achromatism.
- According to the condition for achromatism, the absolute values of the product of the effective focal length and the V-number of the first involved

lens has to correspond to the product of the effective focal length and the V-number of the second one.

- The higher the difference in V-number of the two glasses of an achromatic doublet, the lower the residual aberration.

References

[1] Bentley J and Olson C 2012 *Field Guide to Lens Design* 1st edn (Bellingham, WA: SPIE)
[2] Fischer R E 2008 *Optical System Design* 1st edn (New York: McGraw-Hill)
[3] Kingslake R 1983 *Optical System Design* 1st edn (Cambridge, MA: Academic)
[4] Smith W J 1992 *Modern Lens Design* 1st edn (New York: McGraw Hill)
[5] Schott A G 2010 Optisches Glas—Datenblätter, data sheet (in German)

IOP Publishing

Lens Design Basics
Optical design problem-solving in theory and practice
Christoph Gerhard

Chapter 4

Aberrations and defects in optical systems

During the design of lenses and optical systems, one has to address a number of possible aberrations and image defects. The underlying causes and sources for the formation of those defects are quite different. An optical component or system completely free of aberrations thus cannot exist in practice. Hence, the goal of optical system design is to find the best solution for a given imaging task. Since this task requires knowledge of the nature and mechanisms of aberrations, the most common errors and defects in optical imaging are presented in this chapter. In addition, suitable systems for the reduction of each aberration are introduced.

4.1 Seidel aberrations

4.1.1 Calculation of Seidel coefficients

The so-called **Seidel aberrations** are the five basic monochromatic aberrations of optical systems, given by spherical aberration, coma, astigmatism, Petzval field curvature and distortion. These aberrations can be described and quantified on the basis of **Seidel coefficients** as introduced by *von Seidel*[1] in 1857. As shown in figure 4.1, the impact of a curved optical interface with the radius of curvature R on the direction of an incoming light ray depends on the index of refraction n behind the interface, the entrance height h of the marginal ray[2] at the interface, and the aperture angle u.

Taking these values, i.e. the radius of curvature[3], the index of refraction, the entrance height and the aperture angle into account, the Seidel coefficient of an optical interface can be calculated according to

---

[1] *Philipp Ludwig von Seidel* (1821–1896), a German mathematician and optician.
[2] The marginal ray is the outermost ray that limits a ray bundle.
[3] In the literature, the parameter c instead of the radius of curvature is sometimes given. This parameter is the reciprocal of the radius of curvature, $c = 1/R$.

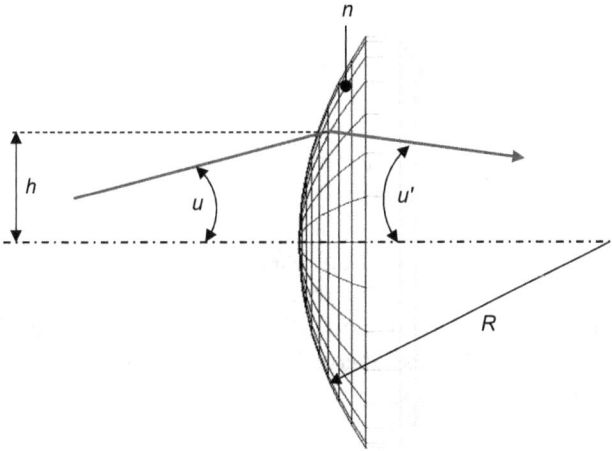

Figure 4.1. Overview of the relevant parameters for the calculation of the Seidel coefficient of an optical interface.

Table 4.1. Seidel sums and corresponding optical aberration.

Seidel sum no.	Optical aberration
I	Spherical aberration
II	Coma
III	Astigmatism
IV	Petzval field curvature
V	Distortion

$$A = n \cdot \left(h \cdot \frac{1}{R} + u \right). \tag{4.1}$$

This calculation is based on the geometric-optical imaging model (see section 2.2) and is principally performed for the marginal ray. However, as we will see in sections 4.1.3–4.1.5, the Seidel coefficient, ray entrance height or aperture angle of the chief ray has to be taken into account additionally in some cases.

The sum of the Seidel coefficients ΣA_i of all the interfaces involved in an optical system (indicated by the index 'i', e.g. six coefficients for a triplet) finally gives the **Seidel sum** as listed in table 4.1. This number or value is a powerful parameter for the quantitative description and evaluation of the imaging quality of optical elements and systems. As presented in more detail in the following sections, each basic Seidel aberration of an optical interface or entire system can be expressed as or determined by Seidel coefficients and Seidel sums. These coefficients or sums, respectively, can be positive or negative values. The lower the absolute value, the better the imaging quality and the lower the particular aberration where very small Seidel sums ($\ll 1$) represent low aberrations.

4.1.2 Spherical aberration

Spherical aberration is a basic optical image defect that is also well known from the human eye [1, 2]. It occurs due to the sphericity of classical optical components such as mirrors or lenses where the surface shape is a spherical segment. As a consequence of this surface shape, a particular angle of incidence results for each ray entrance height h with respect to the optical axis. According to Snell's law, a particular refraction angle thus results for each light ray, depending on its entrance height. As shown in figure 4.2, each light ray of an extended bundle of rays finally features a specific ray entrance height-dependent back focal length (BFL). This behaviour leads to a difference in back focal length ΔBFL between the involved light rays, i.e. the so-called **longitudinal spherical aberration**. This kind of aberration is visualised by the following examples.

Example 1: Spherical abberation of mirrors
The nominal focal length f of a spherical concave mirror with a radius of curvature R is given by the simple interrelation

$$f = \frac{R}{2}. \tag{4.2}$$

However, for extended bundles of light rays, the entrance height h of each light ray has to be taken into account according to

$$f = R \cdot \left(1 - \frac{1}{2 \cdot \cos\left(\arcsin\left(\frac{h}{R}\right)\right)} \right). \tag{4.3}$$

As a consequence, no common focal length can be determined for extended bundles of light rays since the difference in focal lengths Δf amounts to

$$\Delta f = f(h_{\min}) - f(h_{\max}). \tag{4.4}$$

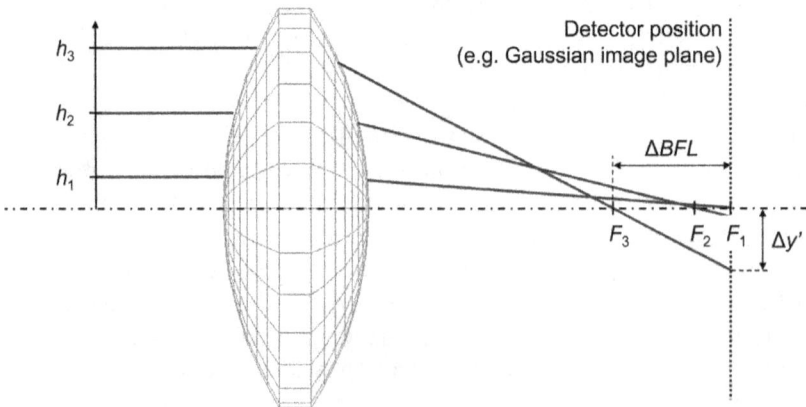

Figure 4.2. Formation of spherical aberration due to different ray entrance heights.

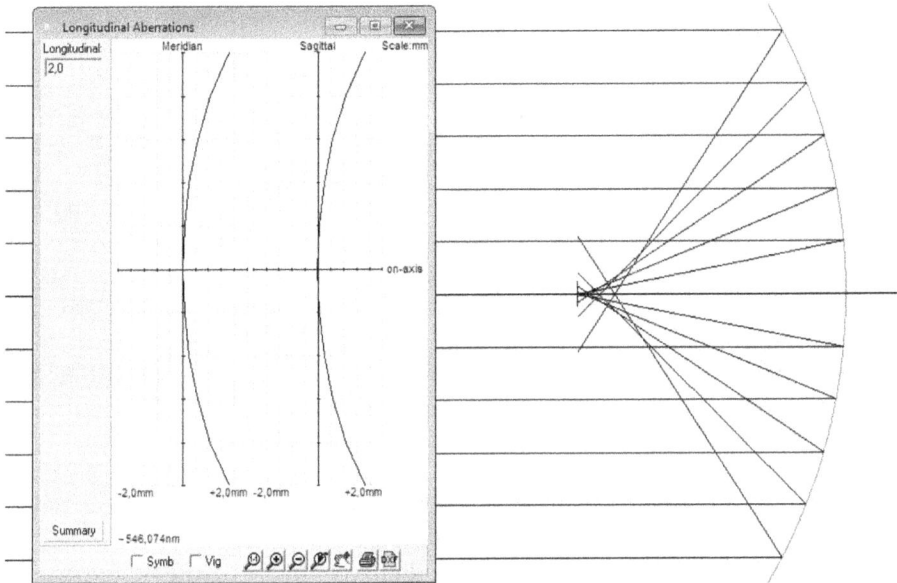

Figure 4.3. Visualisation of the impact of ray entrance height on the focal length of a concave mirror including the longitudinal aberrations diagram.

Here, $f(h_{min})$ is the focal length of the light ray at the minimum entrance height[4] and $f(h_{max})$ is the focal length at the maximum entrance height where the first value is higher than the second one as shown by the comparison in figure 4.3.

> FOCUS ON:
> Spherical aberration of mirrors
> → Exercise **E15**

Example 2: Spherical aberration of lens surfaces
The nominal back focal length BFL of a curved optical interface[5] with the radius of curvature R as, for example, the first surface of a spherical lens can be calculated via the so-called vergence equation given by

$$\text{BFL} = R \cdot \frac{n}{n - 1}. \tag{4.5}$$

Here, n is the index of refraction behind the interface and the index of refraction in front of the interface is assumed to amount to 1. When considering the entrance height of a light ray, equation (4.5) is rewritten as

---

[4] i.e. close to the optical axis.
[5] The back focal length BFL of a single optical interface corresponds to its focal length in the image space f'.

$$BFL = R + \frac{h}{n \cdot \sin\left(\arcsin\left(\frac{h}{R}\right) - \arcsin\left(\frac{h}{n \cdot R}\right)\right)}. \quad (4.6)$$

Equivalent to equation (4.4), the difference in back focal lengths is given by

$$\Delta BFL = BFL(h_{\min}) - BFL(h_{\max}). \quad (4.7)$$

This effect, the longitudinal spherical aberration caused by refraction, is shown by the example of an optical interface between two optical media with the indices of refraction 1 and 1.5, respectively, where the radius of curvature is $R = 100$ mm in figure 4.4.

Apart from longitudinal spherical aberration, which is given by the difference in (back) focal length along the optical axis in propagation direction of light, the effect of **lateral spherical aberration** occurs due to the spherical shape of classical optical surfaces. When taking a closer look at the Gaussian image plane in figure 4.2 it turns out that for this fixed position, which in practice could correspond to the physical position of the detector used for imaging, different image heights y' are found. This secondary error directly follows from longitudinal spherical aberration and gives rise to a blurring of the image.

> **FOCUS ON:**
> Spherical aberration of lenses
> $\rightarrow$ Exercise **E28**

In practice, spherical aberration can be corrected or rather minimised by several actions: first, a reduction of the clear aperture may be sufficient in some cases since shadowing the outer rays of a ray bundle might result in notable enhancement of the imaging quality due to a reduction in aperture angle. Second, an appropriate lens design can be chosen; so-called best form lenses, doublets or even aspherical lenses allow spherical aberration to be minimised. The impact of the basic lens design on the formation of spherical aberration is shown in figure 4.5.

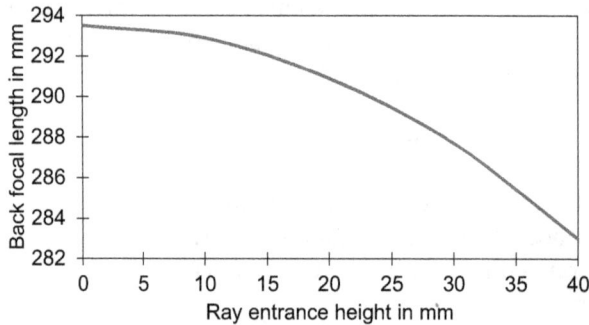

Figure 4.4. Back focal length versus ray entrance height at a spherical optical interface with a radius of curvature of 100 mm; the index of refraction behind the interface is $n = 1.5$.

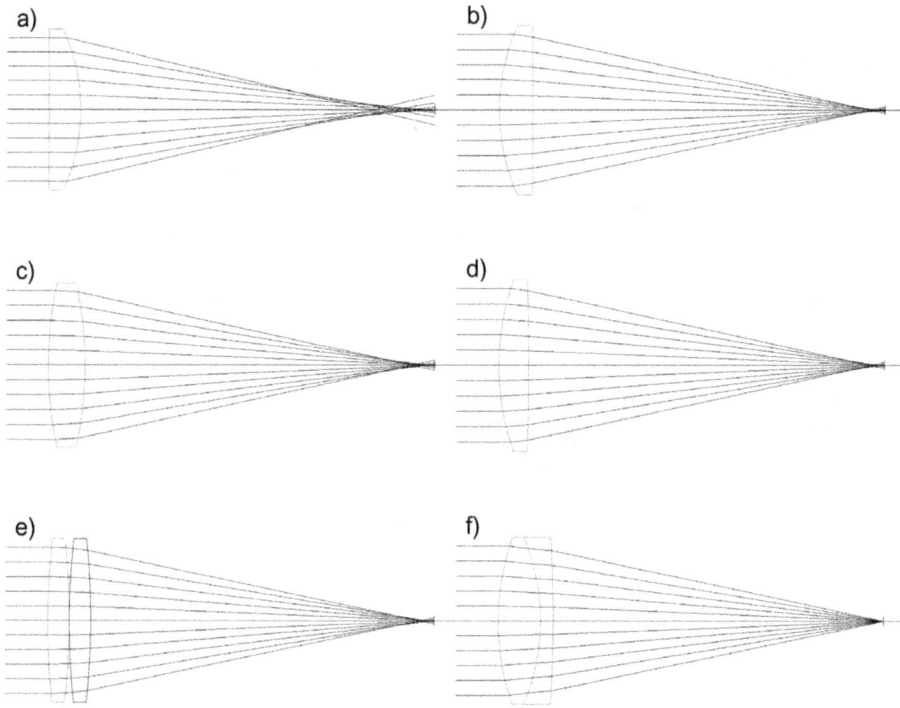

Figure 4.5. Impact of lens design on spherical aberration. Plano-convex lenses feature quite high spherical aberration where the lens orientation has a significant effect (a and b) and best form lenses (d) allow a better correction than symmetric biconvex lenses (c). The best result is found for doublets, either air-gapped or cemented (e and f).

In optical system design, the impact of spherical aberration is quantified by the first Seidel sum, S_I. It is given by

$$S_I = -\sum_{i=1}^{k} A_i^2 \cdot h_i \cdot \Delta_i\left(\frac{u}{n}\right), \tag{4.8}$$

where the parameter A_i follows from

$$A_i = n_i \cdot \left(h_i \cdot \frac{1}{R_i} + u_i\right). \tag{4.9}$$

Moreover, the parameter Δ_i describes the change in aperture angle and index of refraction at an optical interface and can be determined according to

$$\Delta_i\left(\frac{u}{n}\right) = \frac{u_{i+1}}{n_{i+1}} - \frac{u_i}{n_i}. \tag{4.10}$$

It should be noted that in the literature, the reciprocal value $1/R_i$ in equation (4.9) is often expressed by the parameter c_i.

4.1.3 Coma

Spherical aberration as presented in the previous section is a basic image defect that occurs for light rays that propagate parallel to the optical axis. In practice or in most optical sytems, incident light rays are inclined and propagate at a certain angle with respect to the optical axis. As shown in figure 4.6, not only spherical aberration, but additional aberrations occur for such inclined bundles of light.

One of these aberrations is the so-called **coma**, a.k.a. **error of asymmetry**. It appears during imaging of object points outside the Gaussian space, i.e. at high aperture and field angles (>5°) and thus especially for high objects at short object distances. This leads to a notably inclined incidence of light rays on an optical interface and an asymmetric distribution of angles of incidence with respect to the centre axis of the light bundle as shown in figure 4.7.

As a result of such asymmetric distribution of angles of incidence, a lateral shift in image points and a difference in image point positions $\Delta y'$ with respect to the target

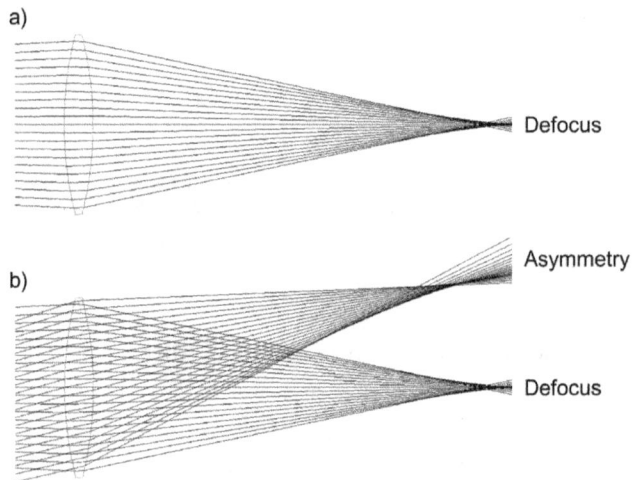

Figure 4.6. Comparison of light propagating on-axis (a) and both on-axis and in the full field (b), i.e. for inclined incidence.

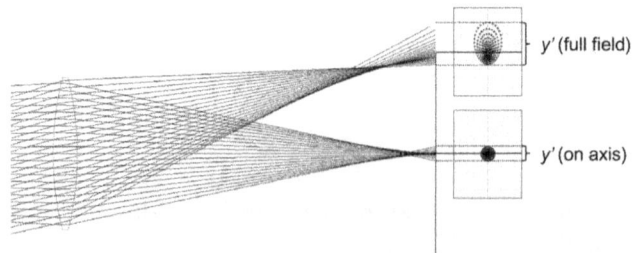

Figure 4.7. Visualisation of the formation of coma by means of the spot diagram.

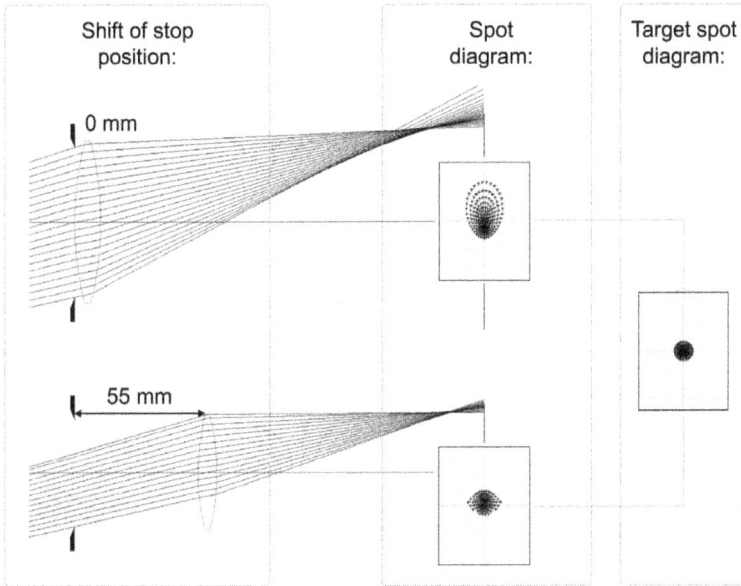

Figure 4.8. Improvement of the image spot size and geometry and reduction in coma, respectively, at the full field angle by shifting the stop towards the natural stop position (by 55 mm in this example).

positions can be observed. Since the resulting image has the appearance of a comet tail, this aberration is referred to as 'coma'. The underlying reason for the formation of coma is the position and diameter of the aperture stop and its distance to the image plane. The symmetric case only occurs for a specific position and diameter where the chief ray represents the centre axis of symmetry in both object and image space as shown in figure 4.8.

Here, all the rays of the ray bundle are nearly focused into the same focal point. This stop position is referred to as the **natural stop position**[6]. As a consequence, coma can be reduced by an appropriate choice of the position of the aperture stop. Another approach is a symmetric general optical system design as, for example, Steinheil-lenses[7] or aplanatic lenses.

The mathematical description and quantification of coma is carried out on the basis of the second Seidel sum S_{II}, given by

$$S_{II} = -\sum_{i=1}^{k} A_i \cdot \bar{A}_i \cdot h_i \cdot \Delta_i\left(\frac{u}{n}\right). \qquad (4.11)$$

---

[6] The natural stop position is sometimes also referred to as the Gleichen's stop position.
[7] Named after the German physicist and optician *Carl August von Steinheil* (1801–1870).

It turns out that in addition to A_i, further Seidel coefficients, $\bar{A}_i$, are considered. The first refer to the marginal ray whereas the latter refer to the chief ray.

> **FOCUS ON:**
>
> Coma
>
> $\rightarrow$ Exercise **E23** and **E28**

4.1.4 Astigmatism

For inclined incidence of rays on an optical interface, different cross sections of the light bundle on the interface and first lens surface, respectively, may occur as shown in figure 4.9. As a result, the footprint of the light bundle on the interface is strongly deformed and non-symmetric. It thus features two different cross sections in X- and Y-direction. These cross sections are referred to as the sagittal and meridional sections. Both sections are arranged perpendicularly to each other and feature different shapes and radii of curvature[8]. Each section features its proper focal plane or point, the sagittal and the meridional one. The result of such imaging is thus the formation of two images

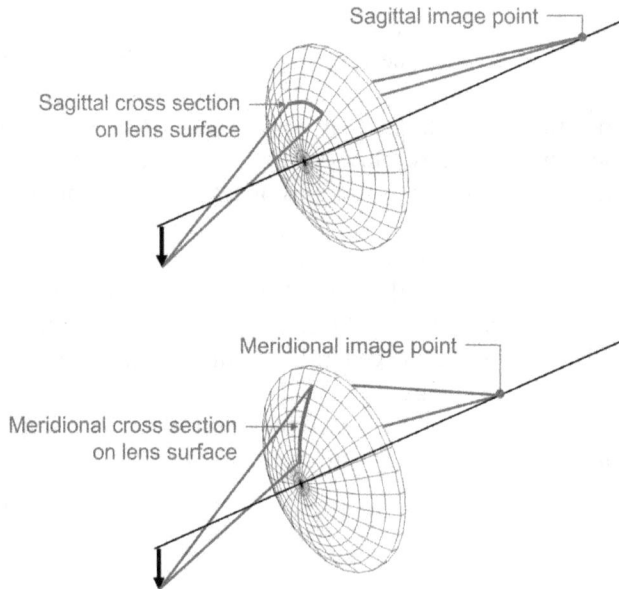

Figure 4.9. Formation of astigmatism due to the difference in image point position for the sagittal and the meridional cross section on the lens surface.

---

[8] As a model for such a surface with different radii of curvature, an old wooden barrel can be imagined: here, the two different radii in horizontal and vertical directions are obvious.

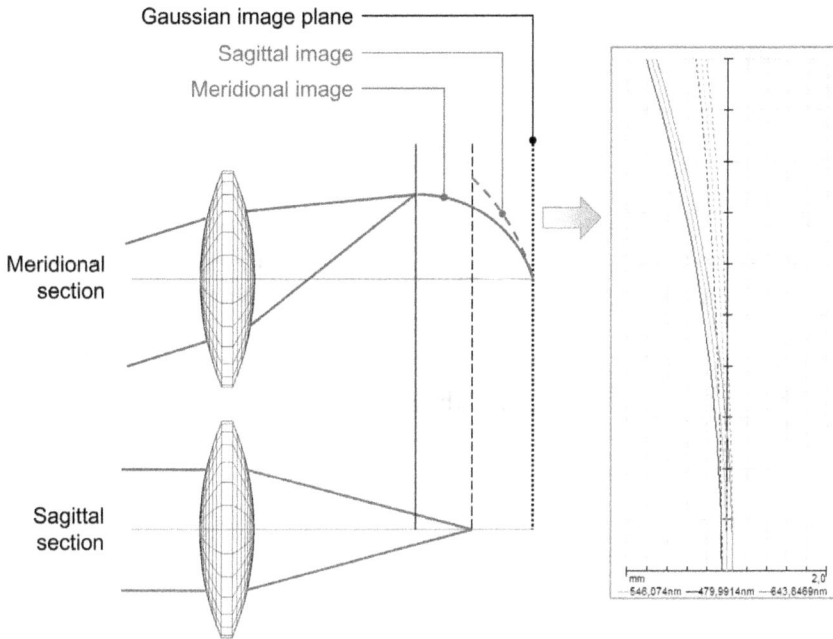

Figure 4.10. Visualisation of two curved images formed in the case of astigmatism (left) including the particular field diagram (right).

behind the focusing optical interface or element—this effect is called **astigmatism**[9]. The two particular focal 'planes' are usually curved[10] and feature different bending as shown in figure 4.10.

As a consequence, no clear image is formed. The best result is found in between the two different focal planes at the so-called **circle of least confusion**. In this medial focus, a blurred image occurs.

This effect or aberration can be minimised by anastigmatic or apochromatic lenses. It is quantified by the third Seidel sum, S_{III}, given by

$$S_{III} = -\sum_{i=1}^{k} \bar{A}_i^2 \cdot h_i \cdot \Delta_i\left(\frac{u}{n}\right). \tag{4.12}$$

4.1.5 Petzval field curvature

Another image defect arising for inclined light rays is **Petzval field curvature**[11]. It preferentially occurs after the correction of astigmatism and can thus be referred to as a residual effect of astigmatic imaging. Here, a rotational-symmetrically curved

---

[9] Astigmatism is a widespread optical defect of the human eye [8].

[10] In professional jargon, curved focal planes are sometimes called 'image bowls'.

[11] This aberration is named after the Hungarian–German mathematician *Jozef Maximilián Petzval* (1807–1891).

image plane is formed; the sagittal and meridional planes thus feature the same radius of curvature, the Petzval radius R_{Petzval}. This radius is given by

$$R_{\text{Petzval}} = \frac{H^2}{n \cdot S_{\text{IV}}}. \tag{4.13}$$

Here, n is the index of refraction behind the optical interface, S_{IV} is the fourth Seidel sum and H is a system-specific parameter. This parameter follows from the entrance height h at the interface and the aperture angle u for both the marginal ray and the chief ray (indicated by overbarred formula signs) according to

$$H = n \cdot (u \cdot \bar{h} - \bar{u} \cdot h). \tag{4.14}$$

The fourth Seidel sum that quantifies the image defect of Petzval field curvature is given by

$$S_{\text{IV}} = -\sum_{i=1}^{k} H_i^2 \cdot \frac{1}{R_i} \cdot \Delta_i\left(\frac{u}{n}\right). \tag{4.15}$$

The parameter $\Delta_i(1/n)$ is calculated on the basis of the involved indices of refraction according to

$$\Delta_i\left(\frac{1}{n}\right) = \frac{1}{n_{i+1}} - \frac{1}{n_i}. \tag{4.16}$$

The effect of Petzval field curvature can be minimised by so-called Petzval lenses[12] or protars[13]. Another approach is the use of curved detectors where the radius of curvature of the detector corresponds to the Petzval radius.

4.1.6 Distortion

Distortion is the last Seidel aberration. It usually occurs for inclined rays in aberration-corrected systems and depends on the position of the aperture stop—similar to coma. Distortion arises due to the fact that the magnification m is not constant over the image plane but either decreases or increases with increasing image height. This leads to a lateral deviation of image point coordinates from the target position that theoretically follows from the nominal magnification[14]. The image of the object geometry is thus distorted. The lateral deviation or difference of the actual image point coordinate y'_a with respect to the target or paraxial image point coordinate y'_p represents the distortion D,

$$D = y'_a - y'_p. \tag{4.17}$$

---

[12] The Petzval lens was the first photographic portrait objective lens and was realised in 1840. It consists of four lenses—a cemented doublet and a split one.

[13] A protar consists of four lenses, arranged in two cemented doublets. It was developed in 1890 and is known as the first anastigmatic lens.

[14] The nominal magnification as calculated based on the object and image heights or distances is only valid for the paraxial case, i.e. small aperture angles.

In practice, the percentaged distortion $D_\%$ is commonly used. This value is given by

$$D_\% = \frac{y'_a - y'_p}{y'_p} \cdot 100\%. \qquad (4.18)$$

We distinguish two different types of distortion: positive distortion, where $V_\% > 0\%$ (and $y'_a > y'_p$) and negative distortion with $V_\% < 0\%$ (and $y'_a < y'_p$). The first case is also referred to as **pincushion distortion** whereas the second case is called **barrel distortion**. This appellation originates from the outer shape of the distorted image as shown in figure 4.11.

> FOCUS ON:
>
> Distortion
>
> $\rightarrow$ Exercise **E29**

The image defect of distortion can be corrected by Steinheil aplanatic lenses, an appropriate choice of the position of the aperture stop or the use of curved detectors. It quantification is carried out via the fifth Seidel sum S_V, given by

$$S_V = -\sum_{i=1}^{k} \left[\frac{\bar{A}_i^3}{A_i} \cdot h_i \cdot \Delta_i\left(\frac{u}{n}\right) + \frac{\bar{A}_i}{A_i} \cdot H_i^2 \cdot \frac{1}{R_i} \cdot \Delta_i\left(\frac{1}{n}\right) \right]. \qquad (4.19)$$

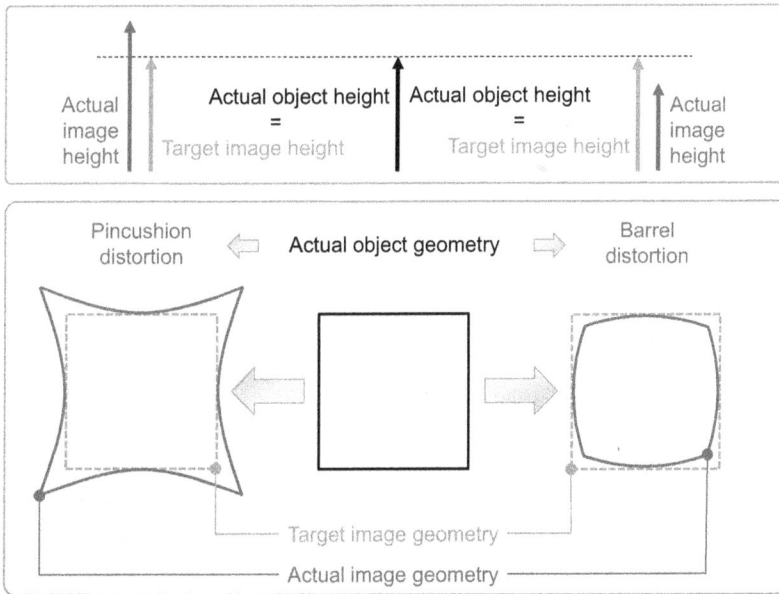

Figure 4.11. The two cases of distortion; pincushion distortion and barrel distortion.

This expression is equivalent to the sum of the third and fourth Seidel sum and can thus be rewritten as

$$S_V = -\sum_{i=1}^{k} \frac{\bar{A}_i}{A_i} \cdot [(S_{III})_i + (S_{IV})_i]. \tag{4.20}$$

Case example 4.1: Characterisation of distortion

Task

The lateral image coordinate of an image is 18 mm. According to the paraxial imaging model this value should amount to 17.5 mm. We are looking for the type and amount of distortion.

Solution

In this case, positive distortion, a.k.a. pincushion distortion, occurs since the value of the actual image coordinate is higher than the nominal one

$$D = 18 \text{ mm} - 17.5 \text{ mm} = 0.5 \text{ mm}.$$

The percentaged distortion amounts to

$$D_\% = \frac{18 \text{ mm} - 17.5 \text{ mm}}{17.5 \text{ mm}} \cdot 100\% = 2.9\%.$$

This value is higher than 0% and confirms the type of distortion as determined above.

4.2 Chromatic aberration

4.2.1 Longitudinal chromatic aberration

Even though **chromatic aberration** is no basic Seidel aberration, it is a common and severe image defect. The underlying reason for chromatic aberration is the intrinsic dispersion of optical media as, for example, glasses (see section 1.3). According to Snell's law, a particular refraction angle consequently results for each wavelength where the refraction angle for 'blue' light (short wavelengths) is higher than for 'red' light (long wavelengths). Thus, different wavelength-dependent foci result for incident white light or light with a broad wave band in general. As shown in figure 4.12, the focal length for 'blue' light is shorter than for 'red' light.

This wavelength-dependency becomes clear when taking a look at the calculation of the back focal length at an optical interface according to equation (4.5) and considering that $n = f(\lambda)$. The resulting difference in back focal lengths for two wavelengths of interest is known as **longitudinal chromatic aberration** ΔBFL.

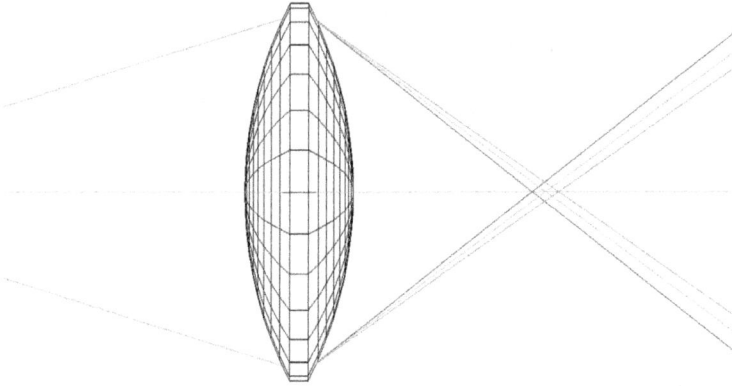

Figure 4.12. Visualisation of chromatic aberration at a single lens.

Case example 4.2: Chromatic aberration of a single lens surface

Task
Behind an optical interface with a radius of curvature of 100 mm the material features the following indices of refraction: $n = 1.96$ at 480 nm and $n = 1.88$ at 700 nm. The index of refraction in front of the interface is 1. Determine the longitudinal chromatic aberration given by the distance between the 'blue' and the 'red' focal point.

Solution
The focal point position with respect to the principal plane of the interface corresponds to the back focal length BFL. It can be calculated by equation (4.5) and amounts to

$$\text{BFL} = 100 \text{ mm} \cdot \frac{1.96}{1.96 - 1} = 204.17 \text{ mm}$$

at a wavelength of 480 nm and

$$\text{BFL} = 100 \text{ mm} \cdot \frac{1.88}{1.88 - 1} = 213.64 \text{ mm}$$

at 700 nm. The difference in back focal length, i.e. the longitudinal chromatic aberration, is thus 213.64 mm − 204.17 mm = 8.47 mm.

Longitudinal chromatic aberration can also be expressed as a function of the nominal effective focal length EFL′ in the image space, the object distance a and the V-number V of the lens material according to

$$\Delta \text{EFL}' = -\frac{a^2 \cdot \text{EFL}'}{(a + \text{EFL})^2 \cdot V}. \tag{4.21}$$

Another possibility is the description via the ratio of the effective focal length and the particular V-number of interest, for example

$$\Delta EFL' = -\frac{EFL'}{V_e} \qquad (4.22)$$

or

$$\Delta EFL' = -\frac{EFL'}{V_d}. \qquad (4.23)$$

This functional notation reveals the basic and logic correlation that the higher the V-number, the lower the amount of longitudinal chromatic aberration.

> **FOCUS ON:**
> Chromatic aberration
> → Exercises **E15**, **E16**, **E17** and **E27**

4.2.2 Lateral chromatic aberration

The formation of longitudinal chromatic aberration gives rise to a chromatic variation of the wavelength-dependent image height. This secondary effect is referred to as **lateral chromatic aberration**. Its level directly depends on the position of the chosen image plane that is also known as the 'mean receiver plane' as shown in figure 4.13.

For two wavelengths, the effect of chromatic aberration can be corrected by achromatic lenses as explained in more detail in section 3.6 and [3, 4]. The two particular wavelengths depend on the choice of the nominal V-number of the lens materials for the converging lens (crown glass) and the diverging one (flint glass).

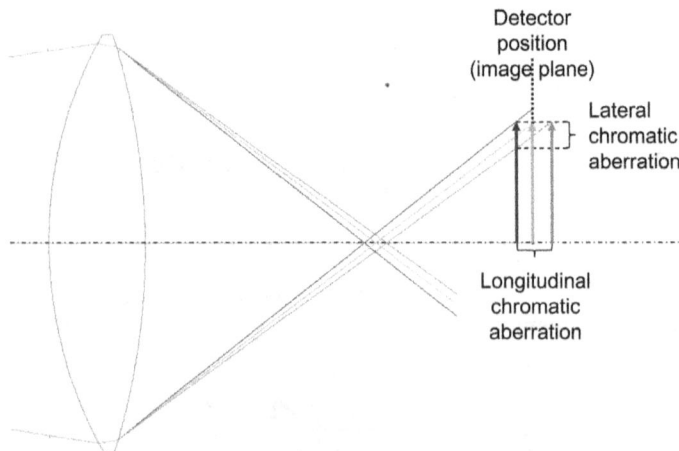

Figure 4.13. Visualisation of the formation of lateral chromatic aberration at a fix detector position.

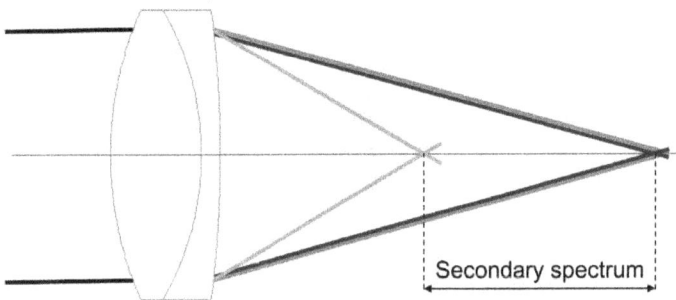

Figure 4.14. Visualisation of the secondary spectrum, i.e. the difference in focal length between two corrected wavelengths (here: blue and red) and a third wavelength (here: green), usually the centre wavelength of the wave band.

For instance, when calculating chromatic aberration based on the nominal V-number V_e, the two corrected wavelengths are given by 479.99 nm (Fraunhofer line F') and 643.85 nm (Fraunhofer line C'), i.e. the wavelengths that define the denominator and thus the main dispersion according to equation (1.8), see section 1.3. Generally, any officially not existing V-number consisting of a customised main dispersion n_x–n_y and centre wavelength n_z may be defined in order to set the specific wavelengths of interest.

After the correction of chromatic aberration for two wavelengths, a residual chromatic error, the so-called **secondary spectrum** remains. As shown in figure 4.14, this error represents the longitudinal chromatic aberration or distance in focal length between the corrected wavelengths on the one hand and the centre wavelength of the chosen wave band on the other hand. This effect results from the different bending of the particular dispersion curves[15] of the involved crown and flint glass. It thus theoretically disappears if the bending is identical for both dispersion curves—which is nearly impossible in practice. For the correction of chromatic aberration for more than two wavelengths, optical systems that consist of at least three lenses are thus required.

4.3 Vignetting

Vignetting is the effect of a decrease in image brightness from its centre to its peripheral sector. As shown in figure 4.15, it is usually found on old photographs and especially portraits and nowadays even caused on purpose for artistic or aesthetic reasons. There are several causes for the formation of vignetting [5] where the most important[16] are mechanical and optical vignetting. Mechanical vignetting occurs in the case of shadowing of marginal light rays that come from the outermost object points by mechanical components such as lens mounts or filter holders. Consequently, the nominal field of view is reduced where the underlying reason is an improper opto-mechanical setup—even though the pure optical setup might not feature vignetting. However, even optical systems without

---

[15] The bending of a dispersion curve is expressed by the partial dispersion, see section 1.3.

[16] 'Most important' in terms of optical system design since both types of vignetting can be considered in the course of calculation and simulation of an imaging system.

Figure 4.15. Example of vignetting in a photograph.

any mechanical restriction can feature pure optical vignetting if the light bundle diameter within the system is limited by one (or more) optical component. This effect is mainly caused by the first elements of an optical setup, leading to a masking of light rays with a high aperture angle.

Comprehension question 4.1: Vignetting

Is there any figure in the present chapter showing the effect of vignetting (except figure 4.15, of course…)?

lens.

coma is reduced, but the bundle of light rays is now limited by the diameter of the

Answer: Vignetting can be observed in Figure 4.8. After shifting the stop position,

---

4.4 Ghost images

Due to reflection or scattering at lens or detector surfaces, mounts, tubes etc, a certain amount of parasitic light is found within the optical path of any optical system. From an energetic point of view, such parasitic light features quite low intensity, but can cause several disturbing effects when being re-reflected onto the detector. First, a notable decrease in image quality can arise by the formation of blurred spots[17] or a reduction in contrast. Moreover, secondary internal images within the optical system—so-called **ghost images**—can be formed.

The evaluation of ghost light paths is a complicated task since there is a considerable number of such paths. For instance, an optical setup consisting of 7 lenses and 14 optical interfaces ($n = 14$), respectively, features 182 ghost light paths ($n^2 - n$). From an energetic point of view, the comparatively high reflectivity of

---

[17] Such blurred spots are commonly referred to as 'orbs'.

detector surfaces[18] can cause intense ghost light paths and in some cases, even damage of optical components may occur[19].

There are three categories of ghost images: first, the classical ghost image where the final image is formed close to—or even directly on—the detector surface. Second, flare where the final image is formed well away from the detector surface, leading to spreading of light and an accompanying general degradation in image quality. Third, internal images that are in fact formed well away from the detector, but directly on a lens surface or within the bulk of an optical component. Ghost images can be avoided or minimised by several actions:

- the use of diaphragms in order to dim out incident light by reducing the stop diameter,
- a blackening of inner mount or tube surfaces,
- the application of antireflective coatings on the surfaces of the involved optical components, and
- an appropriate choice of the radii of curvature of optical interfaces.

4.5 Wave aberrations

Even though the term '**wave aberrations**' may suggest that here, the wave optical imaging model is applied, this type of defect is not described by a straight wave optical consideration. It is rather the determination of the shape and thus deformation of a wave front after transmission through an optical component or system with respect to a reference wave front. This determination is carried out on the basis of pure geometrical calculations. In the simplest case, i.e. for parallel light coming from infinity[20], a plane wave front can be assumed since the light rays (remember: the normal on the wave front) propagate parallel to each other. The flatness of this wave front can be described by several parameters. In optical system design, the root mean squared flatness rms and the peak-to-valley value PV are usually considered. The first parameter, rms, is given by the arithmetic mean of the squares of wave front heights with respect to a mathematically perfect reference flat. In contrast, the PV value describes the maximum difference in height between the highest and the lowest point of a wave front profile. In order to determine the influence of a lens or optical system, wave aberration is measured at the position of the exit pupil[21]. In focusing optical arrangements, spherical waves are emitted by an

---

[18] For instance, the reflectance of CCD-chips amounts to 8%–40% whereas the residual reflectance of a standard coated glass interface is merely 0.1%–0.5%.

[19] In high-power laser systems, light reflected back and focused onto optical surfaces may reach sufficient intensity for laser-induced damage. In order to avoid such damage, optical isolators, a.k.a. Faraday isolators, are implemented into the optical setup.

[20] Sunlight is considered to come from infinity; that means that the object distance is defined to be infinite due to the high distance of the sun. The same approach is used for well-collimated laser irradiation (but is not valid due to the propagation characteristics of laser beams or Gaussian beams in general).

[21] A pupil is a virtual stop that limits the bundle of light rays. It is given by the image of the (mechanical) aperture stop. In a focusing optical setup, there are two pupils, the entrance pupil, i.e. the image of the aperture stop in the object space and the exit pupil, i.e. the image of the aperture stop in the image space. Both pupils are conjugated. An object placed at the position of the entrance pupil is thus imaged to the exit pupil's position by the optical system.

object point according to the Huygens–Fresnel principle (see section 1.2). Both rms and PV are thus determined with respect to an ideal fundamental spherical wave front[22].

> **FOCUS ON:**
> Wave aberration
> $\rightarrow$ Exercises **E31**, **E32**, **E33** and **E34**

4.6 Contrast and contrast transfer

The quality of optical imaging also results from its **contrast**. Contrast transfer is thus of interest, even though is does not represent a classical aberration. Each object features a certain contrast in intensity that is quantitatively given by the **Michelson contrast**[23] C_M, also referred to as **modulation** M as given by:

$$C_M = M = \frac{I_{\max} - I_{\min}}{I_{\max} + I_{\min}}. \tag{4.24}$$

Here, $I_{\max}$ is the intensity of light within bright object areas and $I_{\min}$ is the light intensity at dark object areas [6], for example, the intensities of black and white lines of a black and white stripe pattern. The ratio of the Michelson contrast or modulation of an image and the corresponding object gives the so-called **modulation transfer function** MTF [7]:

$$\text{MTF} = \frac{M_{\text{image}}}{M_{\text{object}}}. \tag{4.25}$$

where the maximum possible value is 1 as exemplified in figure 4.16.

This function depends on the structure of the object, which is considered by the object's **spatial frequency** R, i.e. the reciprocal value of the spatial period length of a structure, given in line pairs or cycles per millimetre. Getting back to the example of a black and white stripe pattern, the spatial period length is given by the distance between black or white lines. The spatial frequency, where no contrast transfer occurs since $\text{MTF}(R) = 0$ is defined as the **cut-off frequency**. It is given by

$$f_{\text{cut-off}} = \frac{1}{\arctan\left(\frac{\lambda}{D}\right)}. \tag{4.26}$$

---

[22] In practice, a wave front and wave front deformation are measured with the aid of a so-called Hartmann–Shack sensor. This device basically consists of a micro lens array and a detector. First, the reference wave front is sectioned into a number of sub-apertures by the micro lens array and the position of the focal point of each sub-aperture on the detector is measured. Second, the wave front is measured again after passing an optical component or system and the displacement of each focal point with respect to its initial position is determined. The wave front deformation caused by the optics is finally calculated from this displacement of focal points.
[23] Named after the American physician *Albert Abraham Michelson* (1852–1931).

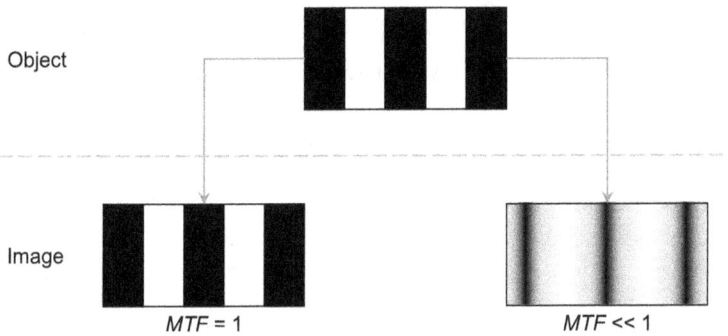

Figure 4.16. Comparison of the particular contrast of an object and corresponding image for high contrast transfer (MTF = 1) and poor contrast transfer (MTF ≪ 1).

Here, λ is the wavelength of light and D is the diameter of the entrance pupil.

Case example 4.3: Determination of the modular transfer function

Task

At a specific spatial frequency of interest, the intensity of light within the brightest areas of an object amounts to 43 W cm^{-2} whereas the light intensity at dark object areas is 21 W cm^{-2}. The image of this object features an intensity of 31 W cm^{-2} at bright image areas and 19 W cm^{-2} at dark image areas. Our goal is now to determine the modular transfer function of the imaging optical system for the given spatial frequency in order to estimate the quality of contrast transfer.

Solution

The Michelson contrast or modulation of the object amounts to

$$M_{\text{object}} = \frac{43 \text{ W cm}^{-2} - 21 \text{ W cm}^{-2}}{43 \text{ W cm}^{-2} + 21 \text{ W cm}^{-2}} = 0.34$$

whereas the modulation of the image is

$$M_{\text{image}} = \frac{31 \text{ W cm}^{-2} - 19 \text{ W cm}^{-2}}{31 \text{ W cm}^{-2} + 19 \text{ W cm}^{-2}} = 0.24.$$

The modular transfer function thus accounts for

$$\text{MTF} = \frac{0.24}{0.34} = 0.71.$$

Hence, moderate contrast transfer by the optical system can be expected.

4.7 Aberrations and defects in optical systems—bottom line

- Seidel aberrations are the five basic monochromatic aberrations of optical components or systems.
- The Seidel aberrations are spherical aberration, coma, astigmatism, Petzval field curvature and distortion.
- Seidel aberrations are quantified by Seidel coefficients and Seidel sums.
- Spherical aberration occurs due to the sphericity of classical lenses where different refraction angles occur for different ray entrance heights.
- Spherical aberration leads to blurring and uneven distribution of brightness in the image plane.
- Coma occurs for imaging of object points outside of the Gaussian space, i.e. at high aperture and field angles.
- Coma is caused by an asymmetric distribution of angles of incidence within a bundle of light rays, leading to a deformation of the image.
- Astigmatism appears due to the fact that a light bundle is subject to different cross sections in the meridional and sagittal plane on a lens surface.
- Astigmatism leads to the formation of different curved focal areas for the meridional and sagittal plane.
- The circle of least confusion is found in between the meridional and sagittal focal area.
- Petzval field curvature describes the formation of a rotational-symmetrical curved image plane.
- Distortion is due to a deviation of the actual magnification from the paraxial magnification.
- Distortion leads to a lateral deviation of actual image point coordinates with respect to paraxial image point coordinates.
- There are two types of distortion of the object geometry—pincushion distortion and barrel distortion.
- Chromatic aberration follows from the wavelength-dependency of the index of refraction.
- For each wavelength, a particular wavelength-dependent focal length is found where the difference between the focal lengths is the longitudinal chromatic aberration.
- Apart from longitudinal chromatic aberration, lateral chromatic aberration occurs during imaging.
- Ghost images occur due to the reflection and scattering of light at lenses, detector surfaces, mounts, or tubes.
- The shadowing of outer rays of a bundle of light rays is called vignetting.
- Wave aberrations describe the deviation of an actual wave front to a theoretical one.
- The ratio of the image and object contrast gives the modulation transfer function.

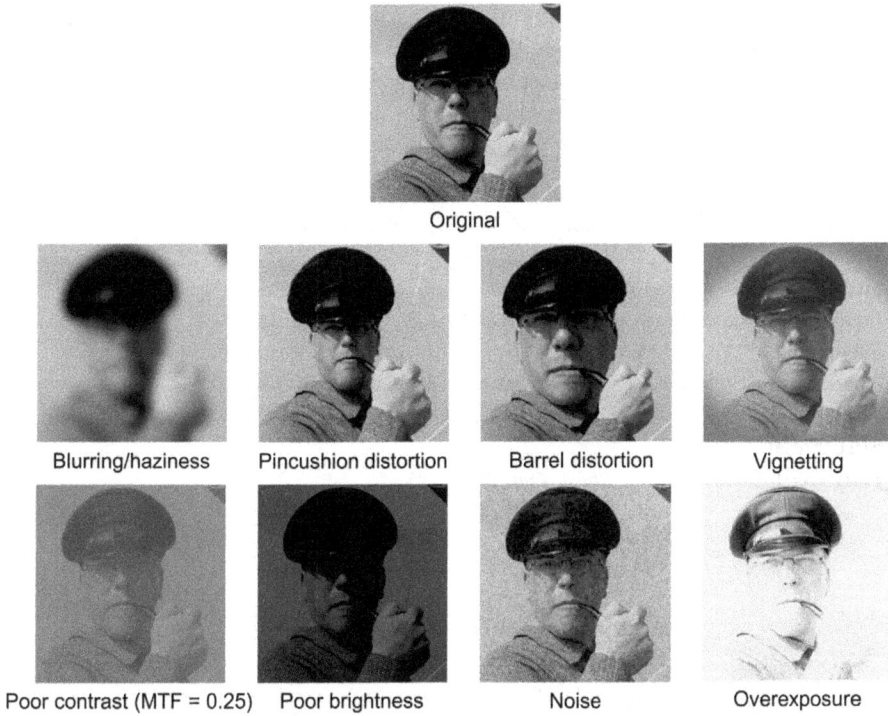

Original

Blurring/haziness   Pincushion distortion   Barrel distortion   Vignetting

Poor contrast (MTF = 0.25)   Poor brightness   Noise   Overexposure

Figure 4.17. Visualisation of impact of aberrations, defects and operator errors on imaging quality.

Spherical aberration	Coma	Astigmatism	Petzval field curvature	Distortion
Asphere	Steinheil lens	Anastigmat	Petzval lens	Steinheil aplanatic lens

Figure 4.18. Overview of optical components and systems for the correction of Seidel aberrations.

- Quite a number of image defects can occur, Seidel aberrations, chromatic aberration and further secondary image defects. Moreover, insufficient cleanliness or transmittance of optics as well as operator errors may further lead to poor imaging quality as summarised in figure 4.17.

- Optical system design means to reduce aberrations of interest with respect to the given task[24].
- There are a number of optical systems for the reduction of each optical aberration as shown by the overview in figure 4.18.

References

[1] Koomen M, Tousey R and Scolnik R 1949 *J. Opt. Soc. Am.* **39** 370–6
[2] Ivanoff A 1956 *J. Opt. Soc. Am.* **46** 901–3
[3] Gerhard C and Adams G 2010 *Imaging Microsc.* **3** 39–40
[4] Gerhard C, Adams G and Wienecke S 2010 *LED Prof. Rev.* **19** 40–3
[5] Ray S F 2002 *Applied Photographic Optics* 3rd edn (Boca Raton, FL: Focal Press)
[6] Michelson A 1995 *Studies in Optics* 1st edn (Mineola, NY: Dover Publications)
[7] Williams C S 2002 *Introduction to the Optical Transfer Function* 1st edn (Bellingham, WA: SPIE Press)
[8] Harris W F 2000 *Ophthal. Physl. Opt.* **20** 11–30

---

[24] As an example, small and perfectly circular focal points should be produced by laser objectives whereas defects arising from the polychromatic character of light can be disregarded in this case.

Chapter 5

Evaluation of imaging performance

In chapter 4, we have seen that quite a number of different aberrations and image defects can occur during imaging of objects by optical interfaces, components or systems. Since the goal of lens design or optical system design is the reduction and minimisation of relevant aberrations, appropriate evaluation tools become crucial in order to visualise and quantify the (residual) grade of aberrations. Here, different evaluation graphs are in hand. Moreover, Seidel coefficients and sums allow an objective assessment of the image quality. In this chapter, the most important graphs and tables used for the evaluation of the imaging quality of an optical component or system are presented and explained.

5.1 Evaluation graphs

Evaluation graphs allow a fast qualitative and quantitative overview on the imaging performance of optical systems. In most optical design software tools, such evaluation can be performed for different wavelengths, wavelength bands, fields of view etc. Moreover, the resolution of graphs can be adjusted by setting the number of traced rays[1]. The most important graphs are introduced in the following sections. The diagrams and plots shown here were taken from the optical simulation software WinLens. In most cases, representation is provided for both the meridional and the sagittal image plane as well as for the on-axis bundle of light rays and the one coming from the outermost object point. The latter case is also referred to as 'full field'.

---

[1] For instance, the number of traced and displayed rays can be defined by the specification of the number of so-called rings (concentric around the optical axis) and fans (spokes that segment the full circle of a lens or stop). A single ray is then traced at each intersection point of those rings and fans.

5.1.1 The spot diagram

The simple **spot diagram**[2] [1] is the least meaningful graph since it merely visualises the shape of image points. This shape is reconstructed by a pattern of traced light rays and the corresponding image height on the detector in—as well as in front of and behind—the paraxial or Gaussian image plane as shown in figure 5.1.

However, spot diagrams can also give valuable information on the distribution of light intensity over the spot size [2]. Moreover, providing a certain experience, the analysis of the spot diagram allows a first estimation of the type of aberrations as shown by figure 5.2. It can be seen that here, the smallest spot diameter is not found at the Gaussian image plane, but located in front of it. This displacement is referred to as **defocus** and can be quantified by adjusting the scaling and the spacing of the displayed slices or sections. In the spot diagram for the full field shown in figure 5.2, an oval or egg-shaped spot can be observed additionally. This behaviour indicates the presence of coma and even the tangential and sagittal coma (for definition see section 6.4.1) can be rated.

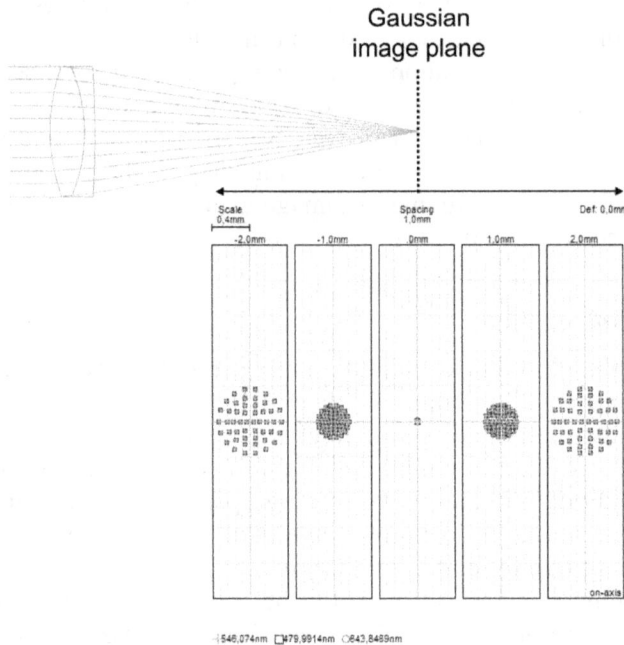

Figure 5.1. Spot diagram for an achromatic doublet showing the image spot at the Gaussian image plane (central spot) as well as one and two millimetres in front of and behind the Gaussian image plane, respectively, for three wavelengths.

---

[2] The software used in this book, WinLens, provides such a simple spot diagram that only shows the position of each traced light ray on the detector, but not the intensity of the particular rays.

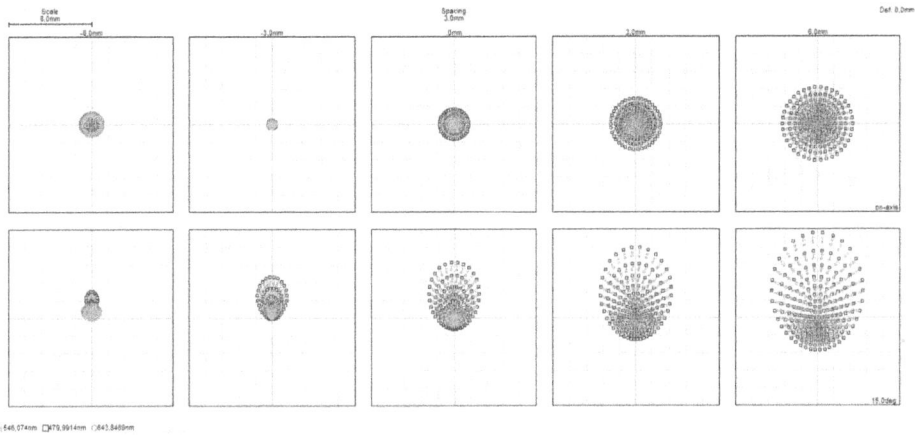

Figure 5.2. Spot diagram showing general defocus (top) as well as coma for the full field (bottom).

Comprehension question 5.1: Spot diagrams

We assume that a spot diagram shows an oval image point in front of the Gaussian image plane. Behind the Gaussian image plane, a second oval image point is observed. The second oval is rotated by 90° with respect to the first one. Which aberration is indicated by these ovals?

Answer: The described behaviour can be explained by astigmatism since here, two image planes, the sagittal and the meridional one are formed. Both planes are well separated and rotated by 90° each other. The section of least aberration, the so-called circle of least confusion, is found in between both image planes.

Moreover, the resolution of an optical system can be ranked. This becomes possible by overlaying the Airy disc that is calculated on the basis of the given parameters (focal length, stop diameter and wavelength, see section 1.7). If the Airy disc diameter is bigger than the spot diameter, the optical system is diffraction-limited and thus features high resolution.

> **FOCUS ON:**
> Evaluating resolution via spot diagrams
> → Exercises **E9**, **E26** and **E34**

Another graphic rendition of image spots is the so-called **full field spot diagram** where the spot patterns of traced light rays are displayed in the form of a quadrant of rotational-symmetric distribution with respect to the optical axis. This representation, shown in figure 5.3, allows the observation of a deformation of the entire image area. It is thus suitable for the identification of distortion or other field aberrations.

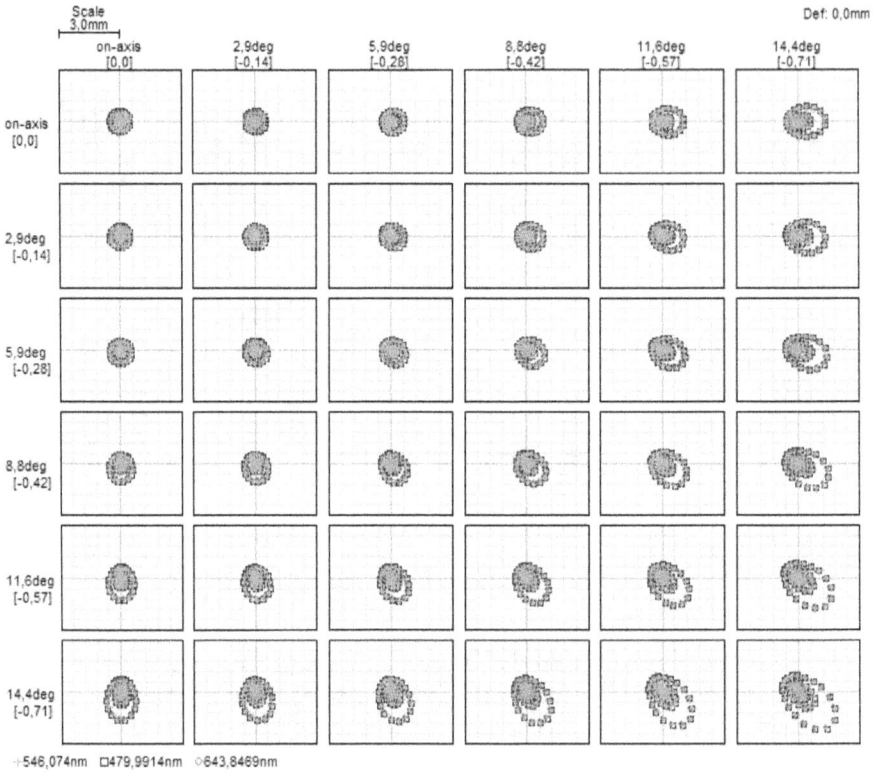

Figure 5.3. Full field spot diagram of a single lens; the defined grid consists of six field points for both fields, the horizontal and the vertical one.

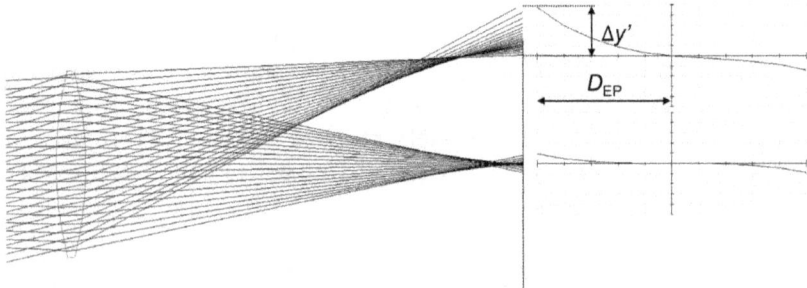

Figure 5.4. Visual description of the transverse ray aberration (TRA) diagram.

5.1.2 The transverse ray aberration diagram

The **transverse ray aberration diagram** or TRA diagram displays the image height as a function of the diameter of the entrance pupil D_{EP} [3] as shown in figure 5.4. It visualises the variation of the image height and is thus an indicator for distortion, defocus and coma.

Comprehension question 5.2: Transverse ray aberration diagrams

What does the transverse ray aberration diagram of a perfect lens or optical system without any distortion, defocus or coma, look like?

Answer: In the ray aberration diagram of a perfect lens or system the resulting graph is given by a straight line that superimposes the x-axis of the diagram.

5.1.3 Field diagrams

Field-dependent aberrations can be evaluated by so-called **field diagrams** [4]. Here, the particular value of aberration is shown on the X-axis and the field angle is shown on the Y-axis where the maximum Y-value is given by the full field of the current imaging case. Four aberrations are covered by field diagrams as shown in figure 5.5: astigmatism, Petzval field curvature, distortion and lateral chromatic aberration.

The field diagram for astigmatism displays the field angle-dependent curvature of the image plane due to astigmatism. In figure 5.5(a), the solid lines represent the meridional plane and the dotted lines show the sagittal plane. Moreover, the diagram includes the maximum deviation φ_{max} between the paraxial image plane and the actual curved image plane that is found for the highest field angle, see figure 5.5(a). As explained in more detail in section 6.2.1, its value can be calculated on the basis of the third Seidel sum. The field diagram for astigmatism also provides additional

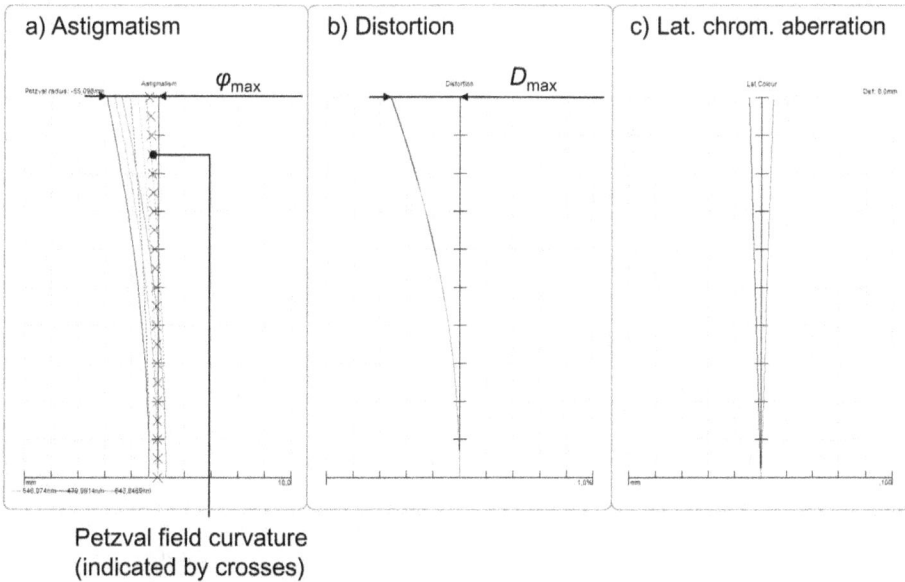

Petzval field curvature
(indicated by crosses)

Figure 5.5. Field diagrams for astigmatism and Petzval field curvature (a), distortion (b) and lateral chromatic aberration (c), including the maximum deviation between the paraxial image plane and the actual curved image plane φ_{max} caused by astigmatism as well as the maximum distortion D_{max}.

information on Petzval field curvature. Its shape is given by the crosses displayed in figure 5.5(a). Further, the actual value of the Petzval radius is usually displayed.

The field diagram for distortion shown in figure 5.5(b) visualises the field angle-dependent distortion, which is either positive or negative. It also includes the maximum deviation in actual image heights or the maximum distortion D_{max} that results from differences in magnification as described in more detail in section 4.1.6. This maximum deviation depends on the fifth Seidel sum, compare section 6.2.1.

Comprehension question 5.3: Field diagram for distortion

Determine the type of distortion shown in figure 5.5.

Answer: The curve for distortion shown in Figure 5.5(b) is bent towards the negative sector of the graph (i.e., towards the left side of the Y-axis). The type of distortion is thus negative distortion, also referred to as barrel distortion (see Section 4.1.6).

Finally, the field angle-dependent lateral chromatic aberration is visualised by the corresponding field diagram, see figure 5.5(c). It shows the differences in image height that are caused by dispersion of optical media. The Y-axis thus represents the 'mean receiver plane' as defined in section 4.2.

5.1.4 Wave front plots

Wave front plots display the deviation in the shape of a wave front Δw with respect to the corresponding idealised theoretical wave front (either plane or curved, depending on the object distance). The quantification of this deviation is is given by the root mean squared error (rms) or the peak-to-valley value (PV) as introduced in section 4.5. There are two main types of wave front plots: the two-dimensional plot is also referred to as optical path difference (OPD) plot or **wave aberration plot** and shows the wave front deviation Δw as a function of the diameter of the entrance pupil D_{EP}. An example is presented in figure 5.6.

The three-dimensional wave front plot shows a topographic map of the wave front surface. As shown by figure 5.7, it further gives the values for the rms and PV deviation of the wave front.

5.1.5 Modular transfer function diagrams

The basic **modular transfer function diagram**, or MTF diagram, displays the MTF (see section 4.6) as a function of the spatial frequency, i.e. the fineness of an abject structure given in line pairs per millimetre. It further quantifies the cut-off frequency where MTF = zero—and thus no contrast is transferred by the optics—as shown in figure 5.8(a). This diagram thus visualises and quantifies the resolution power of an optical component or system. For comparison, the diffraction-limited MTF curve is usually displayed as a benchmark. The MTF can also be represented for a fix

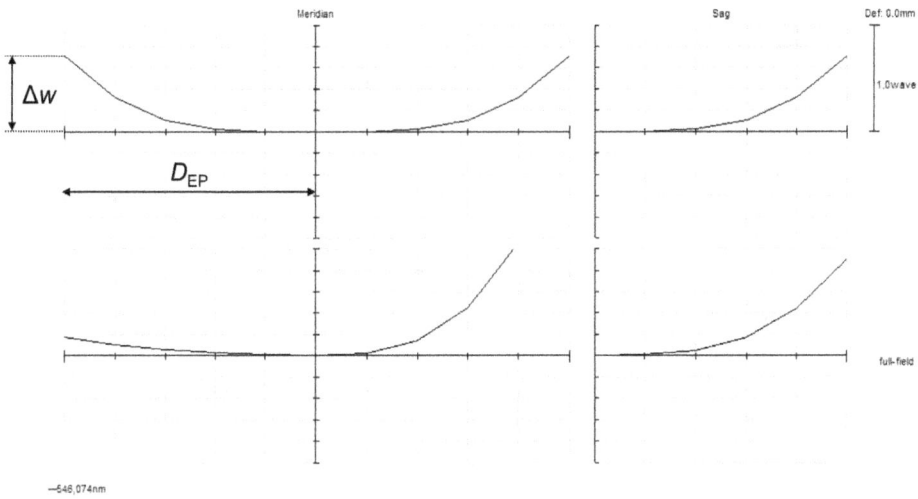

Figure 5.6. Visual description of the 2D wave front plot, a.k.a. optical path difference (OPD) diagram.

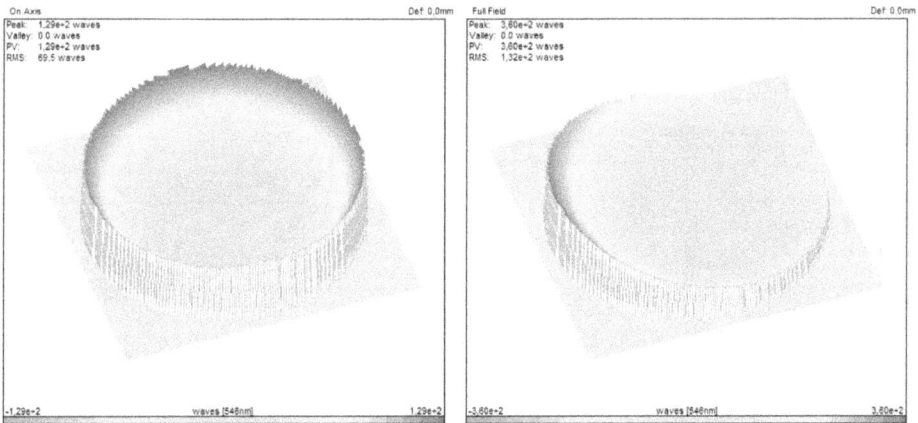

Figure 5.7. 3D wave front plot for the on-axis bundle of light rays (left) and the full field (right) including the rms and PV values for wave front distortion.

pre-defined spatial frequency. Here, it is plotted as a function of defocus, i.e. the distance from the Gaussian image plane as shown in figure 5.8(b). This graph is called the **Through MTF diagram**.

5.1.6 The longitudinal aberration diagram

This diagram displays spherical aberration by plotting the position of the focal point (on the *x*-axis) with respect to the paraxial image plane as a function of the ray entrance height (shown on the *y*-axis). As shown in figure 5.9, the **longitudinal aberration diagram** also visualises longitudinal chromatic aberration when considering more than one wavelength.

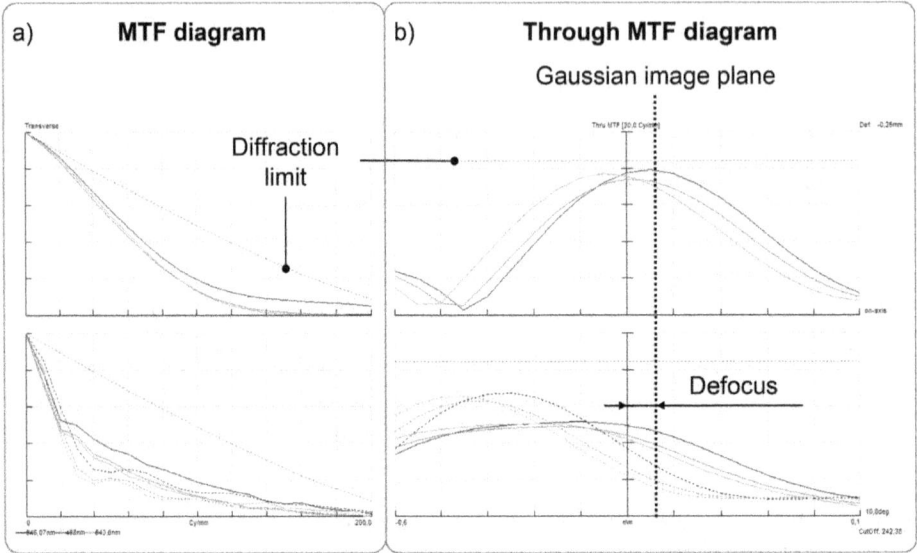

Figure 5.8. (a) MTF diagram and (b) Through MTF diagram for on-axis light (top) and the full field (bottom) at three wavelengths including the diffraction limit.

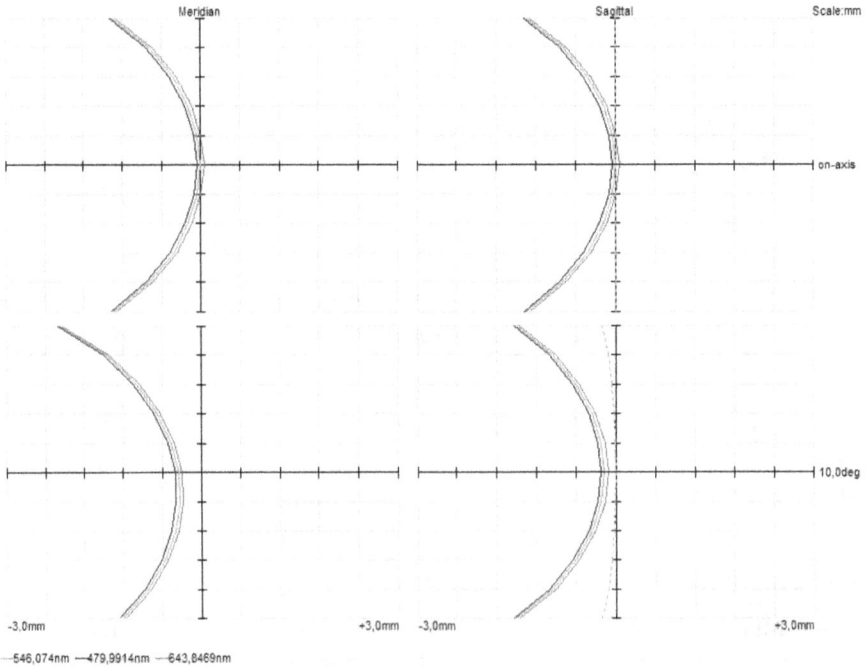

Figure 5.9. Longitudinal aberration diagram for three wavelengths showing spherical aberration in the meridional (left) and sagittal (right) image plane.

5.1.7 The chromatic aberration diagram

Another visualisation of longitudinal chromatic aberration is given by the **chromatic aberration diagram** [5]. As shown in figure 5.10, it displays either the effective focal length (EFL) or the back focal length (BFL) of an optical component or system for each considered wavelength. Moreover, the particular values are provided so the amount of longitudinal chromatic aberration can be determined quite easily.

Comprehension question 5.4: Chromatic aberration diagrams

In figure 5.10, two different chromatic aberration diagrams are shown. The one for the single lens shows a logical behaviour: the shorter the wavelength, the shorter the focal length due to the increasing index of refraction. However, what is the underlying reason or effect for the curve of the achromatic doublet?

Answer: The behaviour shown for the achromatic doublet in Figure 5.10 visualises the effect of the secondary spectrum (see Section 4.2). The doublet is corrected for two wavelengths, but features a residual chromatic aberration with respect to a third wavelength.

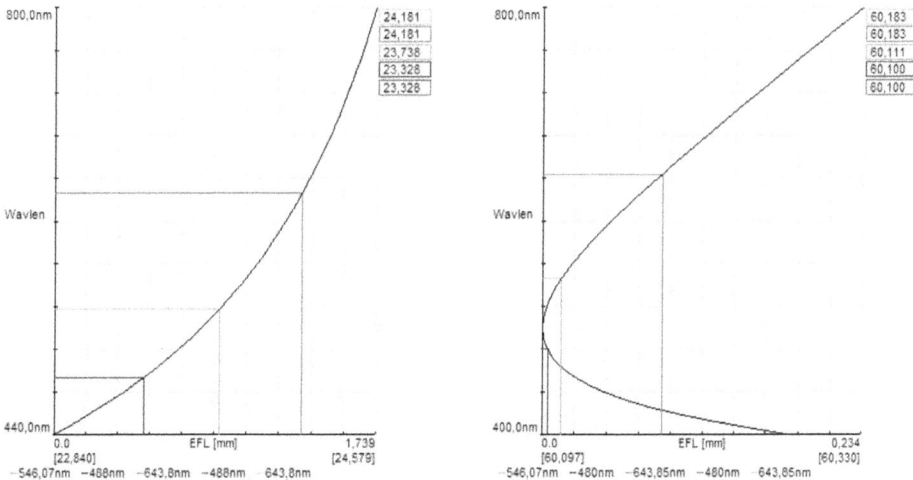

Figure 5.10. Chromatic aberration diagram of a single lens (left) and an achromatic doublet (right) including the particular effective focal lengths for the defined wavelengths (top right).

5.1.8 The Seidel bar chart

As the name implies, the **Seidel bar chart** displays the Seidel coefficient for each surface of an optical system in the form of a bar plot or column chart as shown in figure 5.11. It also includes the Seidel sum where the particular aberrations of interest can be selected easily.

Note that not only the optical interfaces, but also the stop position is considered in this bar chart. The surface of the stop is marked with an asterisk, see the first surface in figure 5.11. The Seidel bar chart is a helpful tool for the identification of critical surfaces, i.e. a surface where the particular aberration preferentially occurs.

FOCUS ON:

Seidel bar chart

→ Exercises **E23**

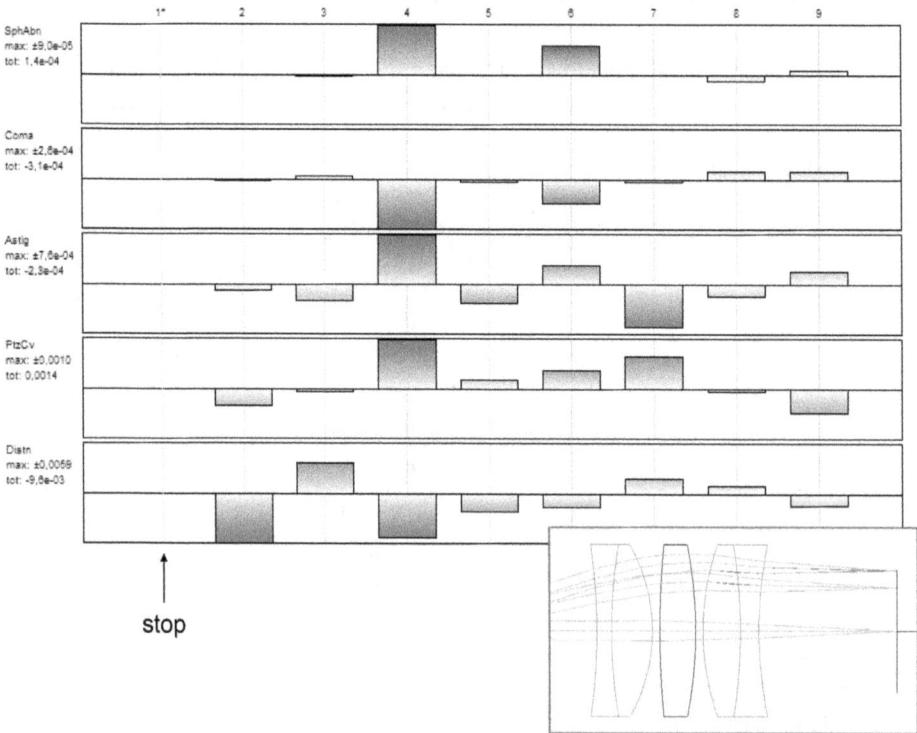

Figure 5.11. Seidel bar chart of an optical system consisting of five lenses—two cemented doublets and a single lens, see inlet—showing (top down) spherical aberration, coma, astigmatism, Petzval field curvature and distortion.

5.1.9 Transmission graphs

The **transmission graphs** show the transmission of an optical system as a function of wavelength. Either the pure transmission caused by internal absorption of the optical media or the total transmission including reflection losses at the surfaces can be visualised. In the latter case, the impact of polarisation of light, a consequence of the Fresnel equations (see section 1.4), can be evaluated. An example of such a transmission graph is shown in figure 5.12.

5.2 Tables and data

The evaluation diagrams presented in the previous sections are based on mathematic models for tracing light rays through an optical system. The relevant numbers and

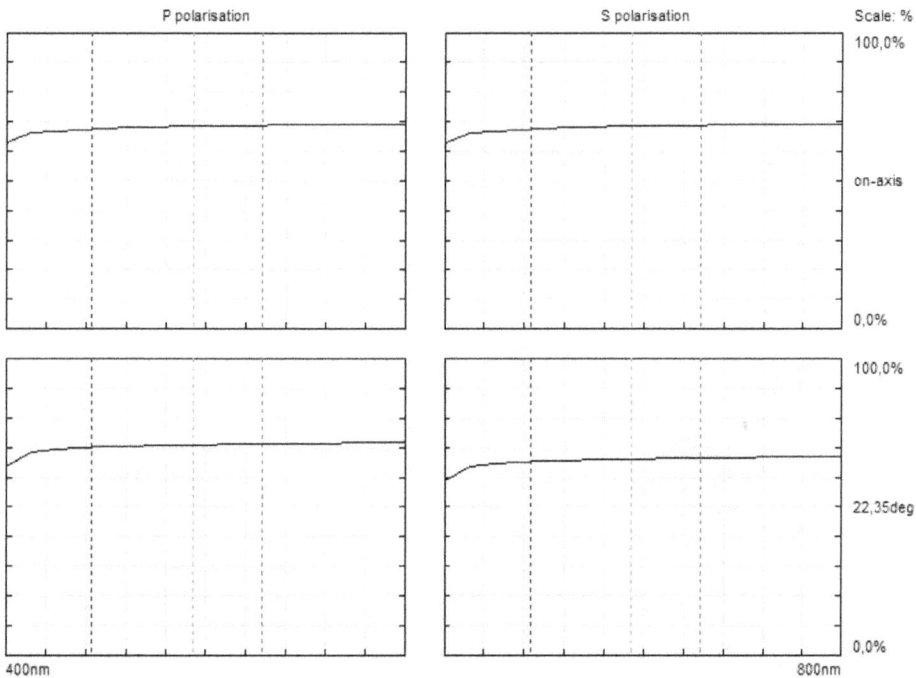

Figure 5.12. Transmission graph of an optical system showing the total transmission for perpendicular polarised light (left) and parallel polarised light (right).

parameters can thus also be accessed and consulted. Such evaluation via tables and data allows a fast qualitative and quantitative overview on the imaging performance of optical systems for different wavelengths, wavelength bands, fields of view etc. Moreover, additional information on optical systems is provided.

For instance, the **surface data table**, shown in figure 5.13, lists information on the physical properties of the optical system, i.e. the radii of curvature and thickness of components, the used glasses etc.

The position of nodal and focal points, total distances and further relevant parameters are summarised in the **paraxial system table**, see figure 5.14, where the given values are calculated based on the paraxial imaging model. This information as well as other values of interest as, for example, the position of the entrance and exit pupil can also be accessed via the **paraxial component table** as shown in figure 5.14.

Main Surface Data | Aspheric coefficients | Diffractives | Tilt/Decenters

	#	Srf	Film	Film wave	Radius	Sepn	Glass		Maker	n	V	Ap/2	Ap
0		Obj				0,00	air	⚏					
1	1	×S			0,0000	12,84	air	⚏					0,80
2	2	S			-47,3150	1,60	SF10	⚏	Schott	1,73	28,40	7,00	fixed
3		S			41,8670	4,80	SK2	⚏	Schott	1,61	56,63	8,00	fixed
4		S			-14,0240	0,50	air	⚏				8,00	fixed
5	3	S			73,9180	4,00	SK2	⚏	Schott	1,61	56,63	9,50	fixed
6		S			-37,0470	0,60	air	⚏				9,50	fixed
7	4	S			21,4410	4,50	SK2	⚏	Schott	1,61	56,63	9,50	fixed
8		S			-45,6430	1,80	SF10	⚏	Schott	1,73	28,40	9,50	fixed
9		S			33,4970	9,98	air	⚏				9,40	fixed
10		Img			0,0000	0,00	mm · def					6,50	

☐ Apply solve for EFL   0,0   target value for EFL

Figure 5.13. The surface data table providing information on physical parameters of an optical setup.

Object side: data wrt FIRST surf		Image side: data wrt LAST surf	
Object distance	0,000	Image distance	9,983
Efl f = PF	-15,811	Efl f' = P'F'	15,811
Front Focus: F	1,562	Rear Focus: F'	9,983
P Plane 1: P	17,374	P Plane 2: P'	-5,828
Nodal Point: N = F-f	17,374	Nodal Point: N' = F'-f	-5,828
Entrance Pupil	0,000	Exit Pupil	169,984
First - Last surf	30,641	First - image surf	40,624
Object height	0,000	Image height	-6,500
Lagrange Invariant	0,329		0,000

	#	Surf	Object distance	Image distance	f	f'	F	F'	P	P'	N	N'	Entr P	Exit P
			0,00	0,00	0,00	0,00	0,00	0,00	0,00	0,00	0,00	0,00	0,00	0,00
1	1	·	0,00	0,00	0,00	0,00	0,00	0,00	0,00	0,00	0,00	0,00	0,00	0,00
2	2		0,00	-112,29	64,97	-112,23	64,97	-112,29	0,00	0,00	47,32	-47,32	-12,84	-18,53
3			-113,89	-88,99	598,65	-556,78	598,65	-556,78	0,00	0,00	41,87	41,87	-20,13	-18,11
4			93,79	38,21	-37,11	23,09	-37,11	23,09	0,00	0,00	-14,02	-14,02	-22,91	-37,26
5	3		37,71	46,27	-121,70	195,62	-121,70	195,62	0,00	0,00	73,92	73,92	-37,76	-87,98
6			42,27	18,38	-98,04	60,99	-98,04	60,99	0,00	0,00	-37,05	-37,05	-91,98	-926,03
7	4		17,78	19,00	35,30	56,74	-35,30	56,74	0,00	0,00	21,44	21,44	-926,63	56,99
8			14,50	15,98	607,00	-652,64	607,00	-652,64	0,00	0,00	-45,64	-45,64	54,49	64,36
9			14,18	9,98	79,49	-46,00	79,49	-46,00	0,00	0,00	33,50	33,50	62,56	169,98

Figure 5.14. Paraxial system table (top) and paraxial component table (bottom) of an optical system.

The ray entrance height and the aperture angle are found in the **paraxial ray trace table**. These parameters are required for the calculation of Seidel coefficients and sums that can be accessed via the **Seidel aberration table** as shown in figure 5.15.

The transmission of each optical interface or component is provided by the **transmission table** as shown in figure 5.16. Such listing is quite helpful in some cases where the particular transmission of a single interface or wavelength is of interest, for example, for the choice of optical coatings. In the transmission table, both internal absorption and reflection losses can be evaluated independently from each other.

Finally, the path of light of an optical system can also be calculated via the Gaussian beam propagation approach [6, 7]. This mathematical method is applied for laser beams and takes the wave physics of light into account. It allows for the

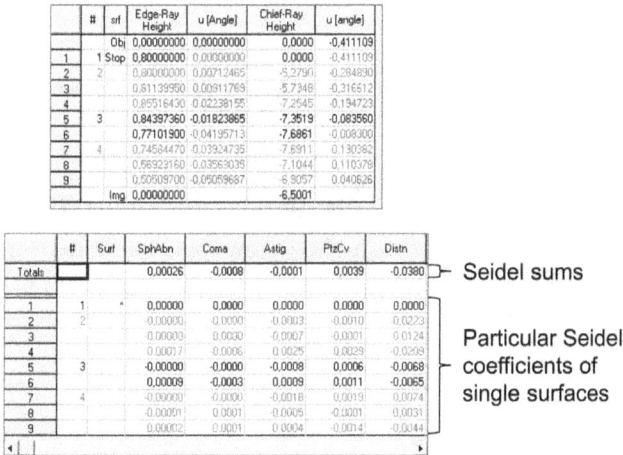

#	srf	Edge-Ray Height	u [Angle]	Chief-Ray Height	u [angle]
	Obj	0,00000000	0,00000000	0,0000	-0,411109
1	1 Stop	0,80000000	0,00000000	0,0000	-0,411109
2	2	0,80000000	0,00712465	-5,2790	-0,284890
3		0,81139950	0,00911769	-5,7348	-0,316612
4		0,85516430	0,02238155	-7,2545	-0,194723
5	3	0,84397360	-0,01823865	-7,3519	-0,083560
6		0,77101900	-0,04195713	-7,6861	-0,008300
7	4	0,74584470	0,03924735	-7,6911	0,130382
8		0,56923160	0,03563039	-7,1044	0,110378
9		0,50509700	-0,05059687	-6,9057	0,040626
	Img	0,00000000		-6,5001	

#	Surf	SphAbn	Coma	Astig	PtzCv	Distn	
Totals		0,00026	-0,0008	-0,0001	0,0039	-0,0380	Seidel sums
1		0,00000	0,0000	0,0000	0,0000	0,0000	
2	2	-0,00000	-0,0000	-0,0003	-0,0010	0,0223	
3		-0,00000	0,0000	-0,0007	-0,0001	0,0124	
4		0,00017	-0,0006	0,0025	0,0029	-0,0209	Particular Seidel coefficients of single surfaces
5	3	-0,00000	-0,0000	-0,0008	0,0006	-0,0068	
6		0,00009	-0,0003	0,0009	0,0011	-0,0065	
7	4	-0,00000	-0,0000	-0,0018	0,0019	-0,0074	
8		-0,00001	0,0001	-0,0005	-0,0001	0,0031	
9		0,00002	0,0001	0,0004	-0,0014	-0,0044	

Figure 5.15. Paraxial ray trace table (top) and Seidel aberration table (bottom) of an optical system.

#	Srf	Film	Glass	587,6nm [custom]	334,1nm	365,015nm	404,656nm	435,834nm	479,991nm	486,133nm	546,074nm	587,562nm	589,294nm	632,8nm
Total				68,59	0,00	16,54	64,11	66,61	67,46	67,56	68,35	68,59	68,60	68,78
1*	1	...		100,00	100,00	100,00	100,00	100,00	100,00	100,00	100,00	100,00	100,00	100,00
1			air	100,00	100,00	100,00	100,00	100,00	100,00	100,00	100,00	100,00	100,00	100,00
2	1	...		92,67	91,33	91,81	92,19	92,39	92,59	92,61	92,79	92,87	92,88	92,95
2			SF10	99,97	0,00	53,51	98,44	99,74	99,89	99,91	99,97	99,97	99,96	99,95
	2	interface		99,87	99,70	99,76	99,81	99,83	35,84	99,85	99,86	99,87	99,87	99,87
3			SK2	99,90	87,29	97,99	99,61	99,71	99,77	99,78	99,90	99,90	99,90	99,91
	3	...		94,57	94,08	94,21	94,32	94,39	94,46	94,47	94,54	94,57	94,57	94,60
4			air	100,00	100,00	100,00	100,00	100,00	100,00	100,00	100,00	100,00	100,00	100,00
3	1	...		94,57	94,08	94,21	94,32	94,39	94,46	94,47	94,54	94,57	94,57	94,60
5			SK2	99,92	89,38	98,32	99,67	99,76	99,81	99,82	99,92	99,92	99,92	99,92
	2	...		94,57	94,08	94,21	94,32	94,39	94,46	94,47	94,54	94,57	94,57	94,60
6			air	100,00	100,00	100,00	100,00	100,00	100,00	100,00	100,00	100,00	100,00	100,00
4	1	...		94,57	94,08	94,21	94,32	94,39	94,46	94,47	94,54	94,57	94,57	94,60
7			SK2	99,91	88,13	98,12	99,63	99,73	99,78	99,79	99,91	99,91	99,91	99,91
	2	interface		99,87	99,70	99,76	99,81	99,83	99,84	99,85	99,86	99,87	99,87	99,87
8			SF10	99,96	0,00	49,43	98,25	99,71	99,88	99,89	99,96	99,96	99,96	99,95
	3	...		92,87	91,33	91,81	92,19	92,39	92,59	92,61	92,79	92,87	92,88	92,95
9			air	100,00	100,00	100,00	100,00	100,00	100,00	100,00	100,00	100,00	100,00	100,00

Figure 5.16. Transmission table of an optical system.

	#	srf	Beam Dia at Surf [mm]	Beam Waist Posn [mm]	Beam Dia at Waist [mm] after surf	Divergence [mRad]	Rayleigh Range [mm]
Initial			1,00	-1000,00	0,75	1,00	748,16
Final			0,00	10,14	0,01	79,07	0,12
		Obj	0,00	-1000,00	0,75	1,00	748,16
1	1	*	1,25	-1000,00	0,75	1,00	748,16
2	2		1,26	-94,88	0,04	11,68	3,17
3			1,28	-71,54	0,03	14,86	2,11
4			1,35	58,27	0,02	34,56	0,63
5	3		1,33	66,78	0,02	28,31	0,58
6			1,22	42,35	0,01	65,48	0,17
7	4		1,18	43,51	0,01	61,52	0,12
8			0,90	45,00	0,01	55,84	0,14
9			0,80	40,78	0,01	79,07	0,12
		Img	0,00	10,14	0,01	0,00	0,00

☐ Beam Propagation Factor M^2

Figure 5.17. Gaussian beam table listing essential parameters related to laser irradiation.

identification of laser beam diameters, the position of beam waists etc. Those parameters are listed in the **Gaussian beam table**[3] as shown in figure 5.17.

5.3 Evaluation of imaging performance—bottom line

- A number of diagrams and tables allow the qualitative and quantitative analysis of optical aberrations and resolution of optical systems.
- For analysis, different numbers of wavelength and traced rays can be considered.
- The spot diagram shows the size and shape of image points in and close to the Gaussian image plane.
- The transverse ray aberration diagram displays the image height as a function of the diameter of the entrance pupil.
- Field diagrams show field-dependent aberrations, i.e.
  - astigmatism,
  - Petzval field curvature,
  - distortion, and
  - lateral chromatic aberration.
- Wave front diagrams show the deviation between a wave front transferred by an optical system and a theoretically perfect curved or plane wave front.
- The wave front distortion can be displayed in 2D or 3D diagrams.
- The basic modular transfer function (MTF) diagram displays the MTF as a function of the spatial frequency.
- The spatial frequency is the fineness of an object's structure.
- The Through MTF shows the MTF as a function of defocus for a fix spatial frequency.

---

[3] Note that in WinLens, also a Gaussian beam graph showing the path of a laser beam through an optical system is available.

- The longitudinal aberration diagram displays spherical aberration by plotting the actual position of the focal point versus the ray entrance height.
- The chromatic aberration diagram shows the effective or back focal length of an optical component or system over wavelength.
- The Seidel bar chart displays the Seidel coefficient for each surface of an optical system in the form of a bar plot or column chart.
- The Seidel bar chart is used for the identification of critical surfaces.
- The transmission graphs show the transmission of an optical system as a function of wavelength.
- The surface data table lists information on the physical properties of the optical system.
- The paraxial system table and the paraxial component table give the positions of nodal and focal points, total distances and further relevant parameters.
- The ray entrance height and the aperture angle are found in the paraxial ray trace table.
- Seidel coefficients and Seidel sums that can be accessed via the Seidel aberration table.
- The transmission of each optical interface or component is provided by the transmission table.
- The Gaussian beam table lists relevant parameters of laser beams.

References

[1] Stavroudis O N and Feder D P 1954 *J. Opt. Soc. Am.* **44** 163–70
[2] Miyamoto K 1963 *Appl. Opt.* **2** 1247–50
[3] Blandford B and Malaka H 2009 *Proc. SPIE* **9666** ETC1
[4] Cai J, Zhang X, Yang Y and Meng S 2018 *J. Phys. Conf. Ser.* **1065** 122004
[5] Blahnik V, Gaengler D and Kaltenbach J-M 2011 *Proc. SPIE* **8167** 81670G
[6] Alda J 2015 Laser and Gaussian beam propagation and transformation *Encyclopedia of Optical Engineering* 2nd edn (Boca Raton FL: Taylor and Francis)
[7] Laures P 1967 *Appl. Opt.* **6** 747–55

Chapter 6

From task to solution

In short: it is a long way from a given imaging task to the final physical imaging system. As a first step, the imaging task and given parameters are analysed and the desired parameters are determined. Based on this basic consideration, an appropriate start system is chosen, evaluated and—as the case may be—optimised until the desired imaging quality is achieved. Finally, the manufacturing tolerances are determined and documented. These steps including the particular theory and approaches are introduced in the present chapter.

6.1 Basic considerations for lens design

6.1.1 Analysis of given imaging tasks

Optical system design can be understood as the very first step of optics manufacturing since the final product of such design is a manufacturing drawing that includes any relevant information for the production of lenses and systems. This includes the materials used such as glasses, the shape of the lenses as given by the radii of curvature and centre thickness, the size of air gaps as well as the required tolerances of all these parameters [1]. Finding an appropriate solution for a given imaging task is thus not only the calculation of focal lengths and the evaluation of imaging performance, but also the optimisation of given start systems and the consideration of possible deviation in form and position of optical components as well as technical feasibility, i.e. final producibility.

The very first step of any optics design process is the analysis of the given imaging task. This analysis allows us to identify given and desired parameters and to define an appropriate start system, which is subsequently evaluated and optimised in an iterative process.

FOCUS ON:
Definition of start systems
→ Exercises **E3** - **E8** and **E19**

In practice, a given set of parameters varies, but typically the object height and distance, the detector size[1] and the illumination, i.e. the wave band, are known initially. The desired information is thus the conjugated parameters as well as the required focal length of the optical system and its setup—the type of imaging system. The conjugated parameters and the focal length can be calculated by quite simple mathematical rules such as the imaging equation or the formula for the magnification as described in more detail in section 3.4. The choice of an appropriate start system, however, requires some experience—or the assistance of convenient software as shown in Case example 6.1.

Case example 6.1: Analysis of a typical imaging task

Task
The goal is to image a rectangular object on a standard 2/3'-CCD chip. The width and height of the object are 280 and 210 mm, respectively. It has to be placed 1 m away from the optical imaging system and the aperture diameter of the system should amount to 8.7 mm. We thus have to determine the focal length and the type of the required optical system as well as the image distance in order to identify the position of the detector within the optical setup.

Solution
The following parameters are given or can be derived from the given information:
1. The object distance[2] is $a = -1000$ mm.
2. The object height can be determined on the basis of the rectangular geometry. According to the Pythagorean theorem[3] $(a^2 + b^2 = c^2)$, the object's diagonal is 350 mm. Since the optical axis is located right in the centre of the object, the maximum object height is half the diagonal[4] and thus $y = 175$ mm.
3. The image height follows from the given detector chip size. Here, we should be careful since the denomination of the detector is quite misleading. The diameter of a 2/3'-chip is not 2/3' (≈ 16.9 mm) and the image size is thus not

---

[1] The detector size can be the size of a camera chip that defines the maximum image height.
[2] Per default, the object distance is a negative value since the object is placed in front of the principal plane (or on the left and thus negative sector of a coordinate system where the principal plane is the Y-axis). The X-axis is given by the optical axis.
[3] Named after the Greek philosopher *Pythagoras of Samos* (c. 570-c. 495 BC).
[4] Half the diagonal since optical system design always 'thinks' rotational-symmetric with respect to the optical axis.

half this value (≈ 8.5 mm)! The value 2/3' merely indicates the size of the connection thread of such chips. Its true dimensions[5] are 8.8 mm in width and 6.6 mm in height. The amount of half the diagonal of a 2/3'-chip, the image height[6], is thus $y' = -5.5$ mm.

The wave band is not specified, so visible light is assumed. The parameters required can now be determined in different ways as shown by the following two approaches.

Approach 1
The first parameter required is the focal length. It can easily be calculated by solving equation (3.8) for EFL:

$$a = \text{EFL} \cdot \left(1 - \frac{1}{m}\right) \rightarrow \text{EFL} = \frac{a}{\left(1 - \frac{1}{m}\right)}.$$

The missing magnification m can be determined using equation (3.7) since both the object height and the image height are known. It accounts for

$$m = \frac{y'}{y} = \frac{-5.5 \text{ mm}}{175 \text{ mm}} = -0.0314.$$

The required focal length is thus

$$\text{EFL} = \frac{-1000 \text{ mm}}{\left(1 - \frac{1}{-0.0314}\right)} \approx 30.5 \text{ mm}.$$

Moreover, we look for the distance between the optical system[7] and the detector. As a first approach, this value is given by the image distance a'. It can be determined on the basis of the magnification according to

$$m = \frac{a'}{a} \rightarrow a' = m \cdot a = (-0.0314) \cdot (-1000 \text{ mm}) = 31.4 \text{ mm}.$$

The required parameters are now defined. But as is characteristic for a good question, the answer gives rise to new questions: which imaging quality in terms of resolution etc is required? Which aberrations and defects are critical and should be considered? Finally, which optical system is the best solution for the given task, a single lens, a doublet, a triplet, a complex setup? This question will be covered in section 6.1.2.

---

[5] The true size of a detector can be found in the literature or databases.
[6] Note that for real imaging, a negative image height results if a positive object height is given and vice versa. Otherwise, if both parameters feature the same algebraic sign, virtual imaging occurs.
[7] Strictly speaking, the distance between the principal plane of the optical system and the position of the detector.

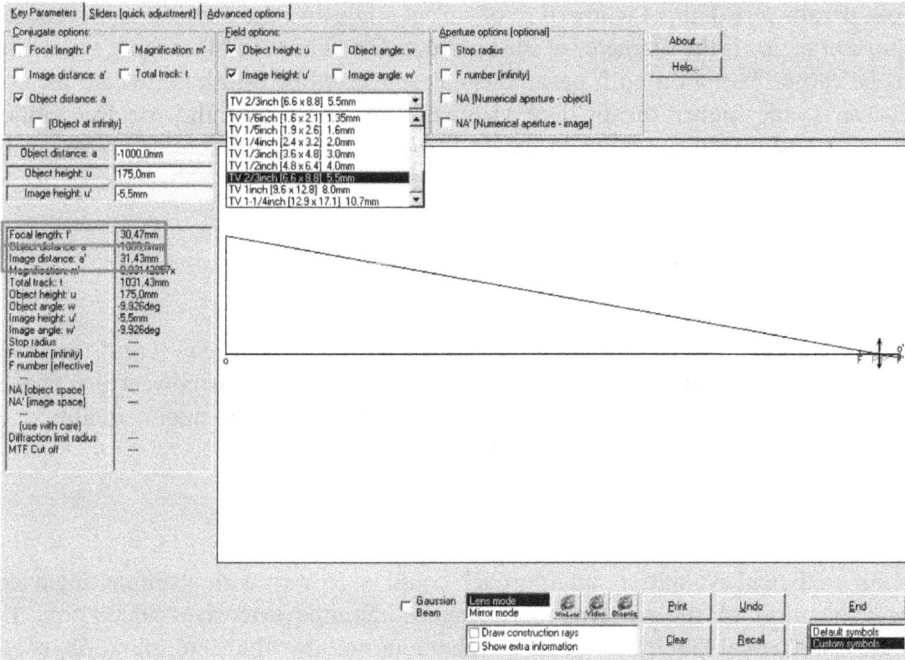

Figure 6.1. Determination of conjugated parameters using the calculation tool PreDesigner; the image height is defined by selecting the detector type.

Approach 2

Nowadays, the required and desired parameters are usually determined with the aid of computer tools and software. Once the basic considerations—i.e. the determination of the object height and the image height, see above—are performed, the conjugated parameters can easily be determined by calculation tools as, for example, the PreDesigner, see figure 6.1.

We see that here, the same focal length and image distance as for the classical or manual calculation presented in approach 1 are found after inserting the object distance as well as the object and image height. Moreover, all parameters that can be calculated based on the given and resulting values are determined. An advantageous feature is the list of detectors provided here: by choosing the detector size or denomination, the corresponding image height is provided automatically.

6.1.2 Definition and choice of start systems

After the calculation and determination of the desired parameters such as focal length, magnification and conjugated parameters, an appropriate start system for the given imaging task has to be chosen. A **start system** is the basic lens setup that promises good imaging quality for a given object and image heights and distances.

Those parameters define the aperture or object angle and the image angle[8]. Another important factor is the diameter of the aperture stop. The type of imaging system can thus be chosen based on both parameters as suggested by Smith [2], where the core message is: the higher the field of view, the more complex the required optical system. Alternatively, an appropriate start system can also be identified by extensive calculations as exemplified in Case example 6.2.

Case example 6.2: Choice of appropriate start systems

Task
In Case example 6.1, we have determined the required focal length and the image distance that result for the given imaging task. The question is now: which type of imaging system, i.e. which arrangement of lenses allows a high imaging quality or an acceptable degree of aberration?

Solution
Approach 1
For an analytical evaluation, an approach could be to start with a simple single lens. The evaluation of the imaging performance requires extensive calculation. For instance, spherical aberration can be characterised by the determination of the particular focal lengths for a certain number of rays at different heights of incidence or distances from the optical axis to the point of incidence, respectively. Longitudinal chromatic aberration is considered by calculating the particular focal lengths for a certain number of rays at different wavelengths within the given spectrum of illumination. Coma, astigmatism and other field-related aberrations are determined by the identification of wavelength-dependent focal lengths and image heights for a certain number of rays and angles of incidence. It turns out that such an evaluation results in an extensive effort. For instance, the calculation of a triplet consisting of six optical interfaces requires 180 single steps if the three basic construction rays and five wavelengths are considered. The alternative approach, computer-assisted optical system design, is thus a time- and cost-saving method as presented hereafter.

Approach 2
The appropriate start system can easily be determined by the use of software tools. In addition to the object distance and height and the image height, information on the free aperture is now needed. This information can be inserted by the clear radius of the aperture or the f-number as shown in figure 6.2.

In the present case, the f-number amounts to

$$\text{f-number} = \frac{\text{EFL}}{D_{\text{aperture}}} = \frac{30.47 \text{ mm}}{8.7 \text{ mm}} = 3.5.$$

---

[8] As a first estimation, those angles can be calculated from simple trigonometrical analysis where the object or image distance is one leg of a right-angled triangle and the aperture radius, i.e. half the aperture stop diameter is the other one.

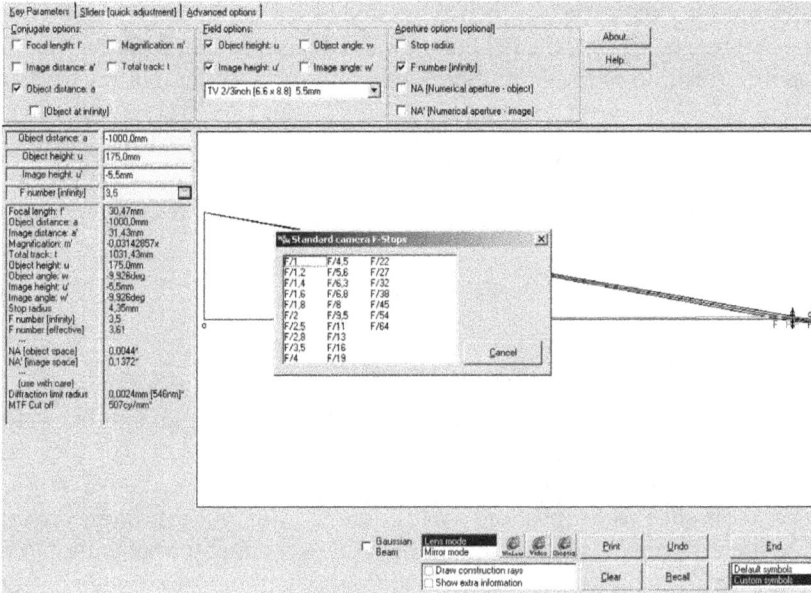

Figure 6.2. Definition of the f-number in PreDesigner.

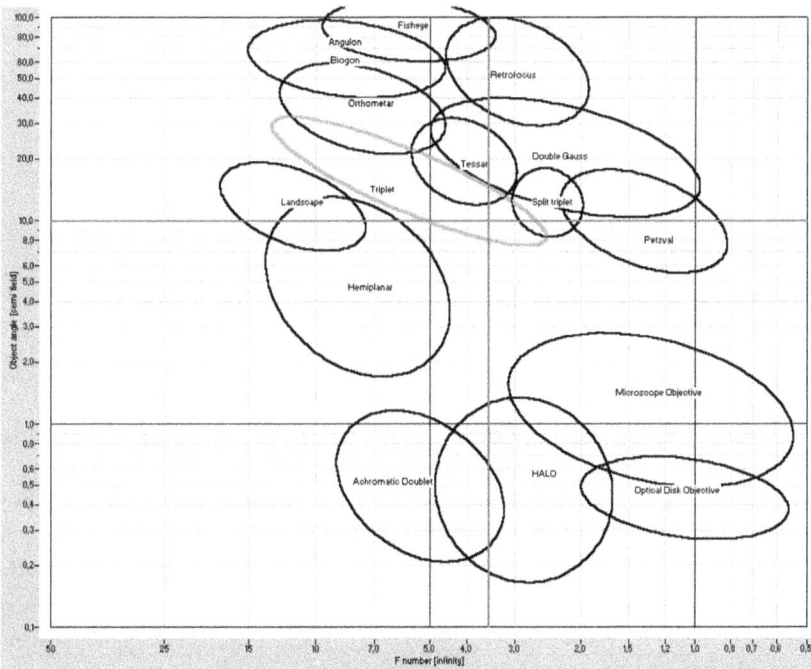

Figure 6.3. Established tool for the selection of appropriate start systems: the f-number versus field plot as suggested in [2].

An appropriate start system is then suggested based on the approach by Smith [2] as shown in figure 6.3. Here, the field of view (FOV), represented by the field angle ω is plotted as a function of the f-number, a.k.a. N or f# and suitable start systems are indicated.

It can finally be stated that based on the given heights and distances and the resulting angles, a triplet with an effective focal length of approximately 30.5 mm turns out to be the most appropriate start system for the given imaging task.

6.1.3 Evaluation of lenses and optical systems

Once the start system is defined the actual optical system design can be performed. This involves the consideration of general conditions[9], the setup of the identified start system as well as its evaluation and potential optimisation. The available evaluation tools such as graphs and tables used for this purpose were presented in chapter 5. Here, special attention should be paid to the target application. You will never find a perfect solution for an optical system without any image defect since optical system design means finding compromises! The goal is thus to minimise relevant aberrations. In special cases, some aberrations or defects could even be ignored as shown by the following examples:

- The field-related defects coma, astigmatism, Petzval field curvature and distortion may be negligible for imaging tasks where large object distances and thus very low aperture angles occur.
- For monochromatic light such as laser irradiation the effect of chromatic aberration does not appear.
- For light coming from infinity[10], nearly no defocus results.

However, it might turn out that a chosen start system should be optimised in order to achieve the required target imaging performance and quality. Such optimisation is presented in the following section.

6.2 Optimisation

Principally, an optical system is not a perfect system addressing and correcting any and every potential aberration or defect. As a result, each optical system can be potentially optimised [3, 4]. Such **optimisation** can be realised by two different main approaches: first, direct optimisation by modifying the physical lens or system parameters and second, indirect optimisation by adding auxiliary correction optics to an existing optical system.

---

[9] General conditions are, for example, the indices of refraction in the object and image space, the considered wavelengths etc.

[10] Actually, an infinite object distance does not exist in practice. But it can be assumed for far-off targets as, for example, cosmic objects such as stars or galaxies.

6.2.1 Direct optimisation via merit functions

Direct optimisation of optical systems is the variation of physical parameters until the residual aberrations of interest are within an acceptable range. In optical system design, these physical parameters are referred to as **variables** and include radii of curvature, surface sphericity or asphericity, centre thicknesses, material data such as indices of refraction, Abbe numbers etc, distances and air gaps, and stop diameters. The aberrations to be corrected are usually called **defects**. Apart from the classical Seidel aberrations (see section 4.1), some other parameters as, for example, paraxial data (conjugated parameters) or aperture and field parameters can be considered as defects. These parameters are not defects in a classical sense, but directly impact optical aberrations as listed in table 6.1.

Knowing the actual defect d_a of an optical system, for example, the value of the Seidel sum of an aberration of interest and the appropriate target value d_t, allows us to define the residual defect value Δd according to

$$\Delta d = d_a - d_t. \tag{6.1}$$

For a single defect, this value thus gives the difference between the current, actual state of the optical system and the requested one. In most cases, a number of quite different aberrations or defects have to be addressed and optimised. Since the defects of interest may have different units, a mix of units and dimensions can result. This means that summing up different defects is problematic or even impossible. In order to overcome this issue, relative defects d_{rel} are used in practice. A relative defect is given by the amount of the ratio of the residual defect value and the acceptable fault tolerance d_{tol},

$$d_{rel} = \left| \frac{d_a - d_t}{d_{tol}} \right| = \left| \frac{\Delta d}{d_{tol}} \right|, \tag{6.2}$$

where the acceptable fault tolerance results from the pre-defined upper and lower defect limit d_{max} and d_{min}, respectively, according to

$$d_{tol} = \frac{d_{max} - d_{min}}{2}. \tag{6.3}$$

Table 6.1. Impact of aperture and field parameters; here: aperture angle and radius of the entrance pupil, on Seidel aberrations.

Seidel aberration	Parameter	
	Aperture angle, u	Radius of the entrance pupil, r
Spherical aberration	—	r^3
Coma	u	r^2
Astigmatism	u^2	r
Petzval field curvature	u^2	r
Distortion	u^3	—

Relative defects defined in that vein are finally used for setting up a so-called **merit function** (MF). This function is the sum of all squared relative defects of the aberrations or defects of interest[11], given by

$$\text{MF} = \sum_i d_{i,\,\text{rel}}^2 = \sum_i \left(\frac{d_{i,\,\text{a}} - d_{i,\,\text{t}}}{d_{i,\,\text{tol}}} \right)^2 = \sum_i \left(\frac{\Delta d_i}{d_{i,\,\text{tol}}} \right)^2. \tag{6.4}$$

A merit function thus finally gives an absolute value. The smaller this value, the better the performance of an optical component or system.

FOCUS ON:
Setting up a merit function
→ Exercises **E24** and **E25**

There are three different types of merit functions: the total MF, the working MF and the concern MF. In the first type, all possible relative defects are taken into account. In a working MF, all activated relative defects, i.e. the defects and aberrations of interest are considered. For the calculation of the concern MF, all relative defects smaller than 1 are excluded, which allows one to define the condition MF = 0 as a stop criterion for calculation. In any case, the goal is to minimise the merit function and the residual aberration, respectively. This is achieved by varying the variables, i.e. the physical lens and system parameters within pre-defined limits. The merit function is calculated using the damped least square (DLS) method[12], i.e. a least squares curve fitting [5]. In this context, the pre-defined limit parameters are of specific interest. Depending on the chosen variable parameters and the acceptable fault tolerance, a certain range within the matrix of possible solutions is specified. This means that merely global minima of merit functions and optima of the optical system's theoretical performance, respectively, can be found by the merit function [6]. As visualised in figure 6.4, much better solutions may be in hand quite close to the defined range, but not identified[13].

One can easily imagine that the definition of various defects, and each corresponding lower and upper limit as well as the variables including reasonable limits[14],

---

[11] The number of aberrations or defects of interest is characterised by the counter index (i) in the merit function.

[12] This approach is also known as the Levenberg–Marquardt algorithm (LMA), named after the American statisticians *Kenneth Levenberg* (1919–1973) and *Donald W. Marquardt* (1929–1997).

[13] This circumstance is sometimes referred to as the 'dilemma of optics designers'.

[14] When defining limits for physical lens parameters technical restrictions must be considered. Even though a virtual glass with an index of refraction of, for example, 650 and an Abbe number of −120 may lead to an excellent theoretical result, it does not exist...and most probably never will. Moreover, frontiers in manufacturing technology should be known and kept in mind during lens design. Finally, pricing is usually an issue, so a free-form lens made of expensive special material may solve imaging problems, but is not a convenient choice for large-scale production of competitive consumer goods.

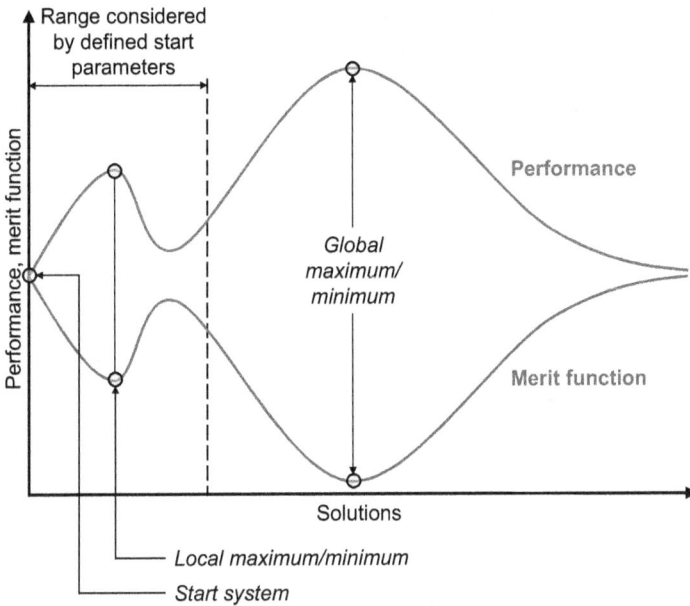

Figure 6.4. Visualisation of the impact of start parameters on the considered range of possible solutions.

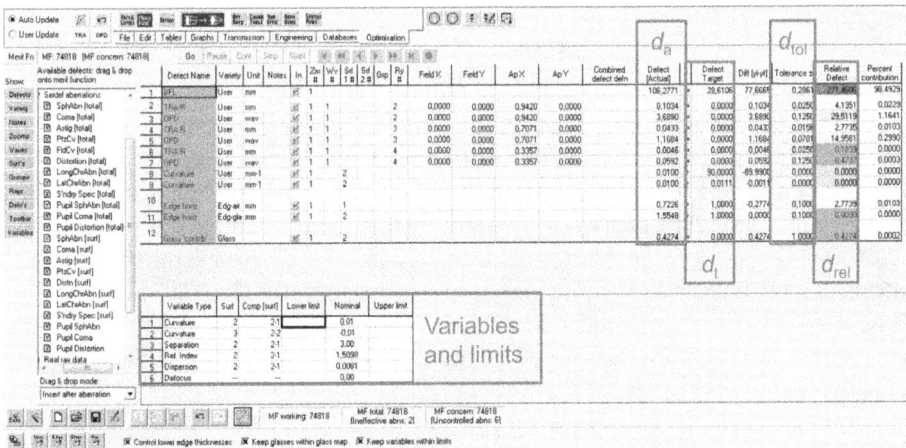

Figure 6.5. Example of the definition of a merit function including the variables and limits, the fault tolerances, actual and target defect values and the relative defect.

leads to a quite extensive matrix. An example of such a definition is shown in figure 6.5.

However, the merit function is a powerful tool for the optimisation and evaluation of the quality and performance of any imaging system. Figure 6.6 shows the result of a case study where an achromatic doublet with a focal length of 50 mm was optimised regarding a special type of aberration, transverse ray aberration.

It can be seen that in this example, the value of the merit function was significantly reduced (figure 6.6(a)). Consequently, both the transverse ray aberration diagram and the spot diagram were noticeably improved (figure 6.6(b) and (c)). This improvement was achieved by a variation of the physical parameters of the achromatic doublets as illustrated in figure 6.6(d). By the alteration of the centre thicknesses and radii of curvature the focus approached the Gaussian image plane, which is equivalent to a reduction of the transverse ray aberration.

> **FOCUS ON:**
> System evaluation via merit functions
> → Exercise **E29**

As an intermediate summary, merit functions are defined based on actual aberrations of an optical system and the acceptable aberrations, i.e. the target defects. The target defect values follow from the particular target functionality and application of an optical system and can be determined with the aid of Seidel sums. For instance, the spot diameter D shown in the spot diagram follows from the Seidel sum for spherical aberration, S_I, according to

$$D = \frac{S_I}{n \cdot u'}. \tag{6.5}$$

Here, n is the index of refraction of the surrounding medium and u' is the image angle[15]. Equating the spot diameter with the pixel size of a given detector and solving equation (6.5) for S_I thus gives the target value d_t for the calculation of the merit function. For that purpose, the actual value d_a is given by the present (start) system and the acceptable fault tolerance d_{tol} can be derived from basic considerations[16]. This basic process also applies to other target defect values:

The defocus φ, i.e. the longitudinal shift of the actual image plane with respect to the Gaussian or paraxial image plane, follows from the first Seidel sum S_I according to

$$\varphi = \frac{3}{8} \cdot \frac{S_I}{n \cdot u^2} \tag{6.6}$$

---

[15] Note that for calculation, angles must be converted into radian where $1° \approx 0.018$ rad.

[16] Unfortunately, the definition of acceptable fault tolerances is not that easy since a number of different parameters and limits have to be taken into account. Apart from theoretical calculations regarding the acceptable range of aberrations and defects, this includes quite different aspects and practical reasons. For instance, the stability of an optical system that—in extreme cases—can also be influenced by changes in temperature and humidity during the operation of the system, the realisable manufacturing accuracy for the production of the involved lenses and mechanical holders, existing patented optical designs from competitors etc, may play a considerable role.

Before optimisation   **After optimisation**

a)

b)

c)

d)

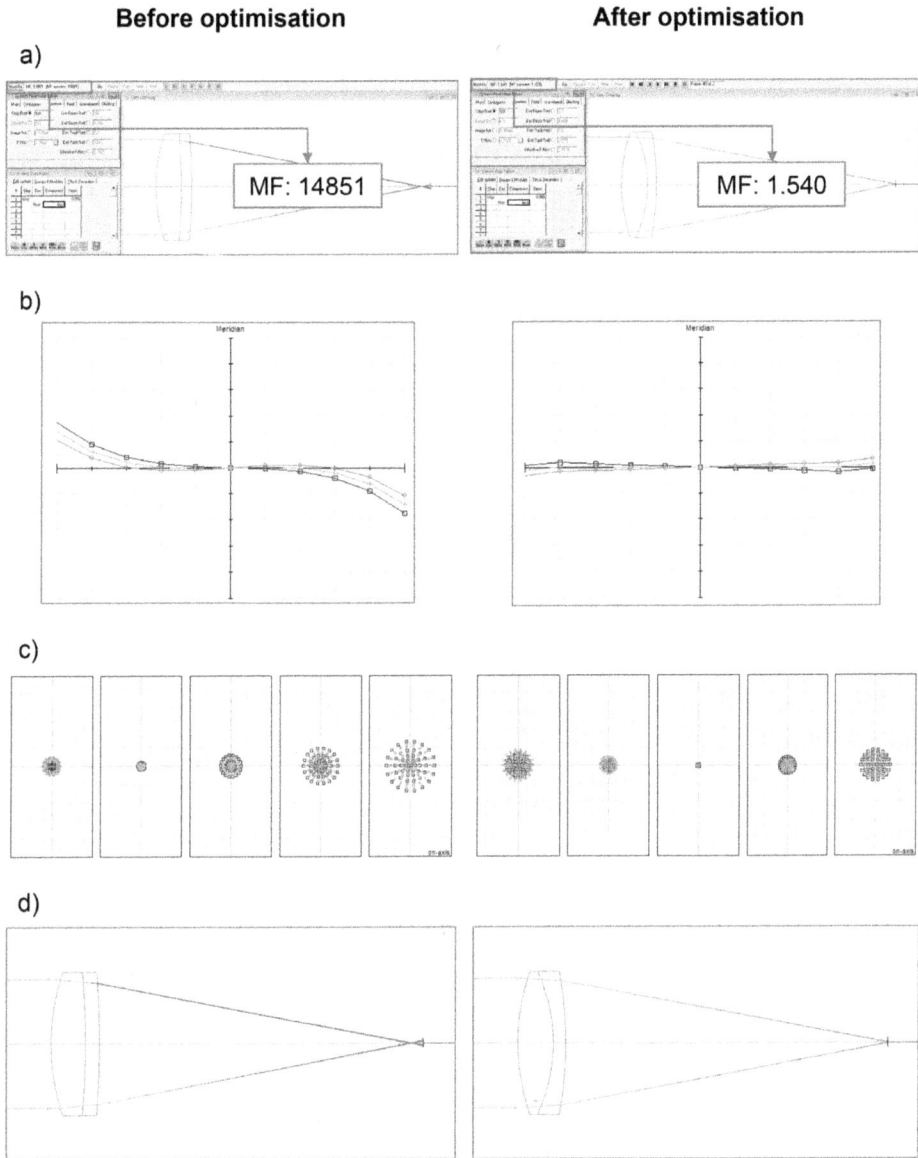

Figure 6.6. Example for the optimisation of an achromatic doublet.

with n being the index of refraction of the surrounding medium and u being the aperture angle.

Coma can be characterised using the second Seidel sum S_{II}. Since the presence of coma can be described as an asymmetric-elliptical deformation of image points, two different lateral spot sizes occur where the maximum lengths are arranged

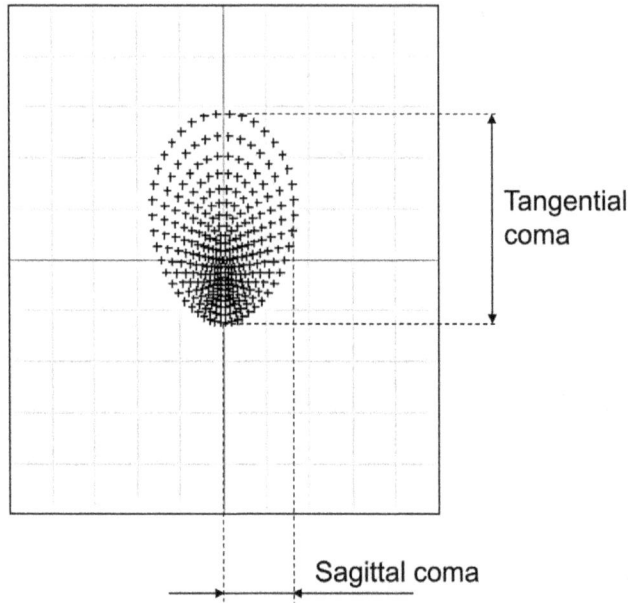

Figure 6.7. Definition of tangential and sagittal coma.

perpendicular to each other. As shown in figure 6.7, these lengths are referred to as tangential and sagittal coma.

Tangential coma $\delta_{\tan}$ and thus the height or maximum extension of the image point is given by

$$\delta_{\tan} = \frac{3}{2} \cdot \frac{S_{\mathrm{II}}}{n \cdot u}. \tag{6.7}$$

Sagittal coma $\delta_{\tan}$ that gives the width of the image point follows from

$$\delta_{\mathrm{sag}} = \frac{1}{2} \cdot \frac{S_{\mathrm{II}}}{n \cdot u}. \tag{6.8}$$

As discussed in more detail in section 4.1.4, astigmatism leads to the formation of two differently curved focal or image sections, the sagittal and the meridional one. The maximum distance between the outermost image section and the Gaussian image plane $\varphi_{\max}$ depends on the third Seidel sum S_{III} and is given by

$$\varphi_{\max} = -\frac{S_{\mathrm{III}}}{n \cdot u^2}. \tag{6.9}$$

The effect of equally curved sagittal and meridional image sections is known as Petzval field curvature. The image 'plane' is then given by a concave surface. Its radius of curvature, the Petzval radius R_{Petzval} follows from the fourth Seidel sum:

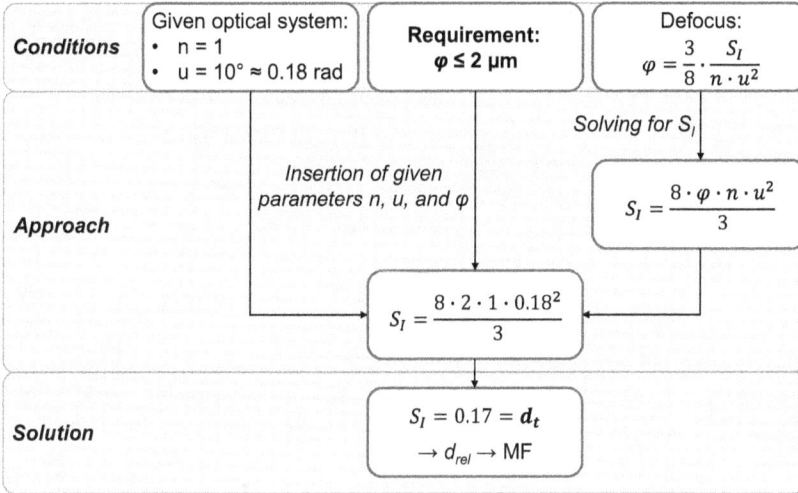

Figure 6.8. Example for the determination of the target defect d_t for setting up a merit function MF based on relative defects d_{rel}. For a given optical system, a limit for the defocus is given as maximum defect. The Seidel sum S_I calculated as shown finally gives the acceptable target defect.

$$R_{\text{Petzval}} = \frac{H^2}{n \cdot S_{\text{IV}}}. \tag{6.10}$$

The parameter H depends on the index of refraction n of the surrounding medium as well as the aperture angle u and the ray entrance height h for both the marginal ray and the chief ray[17] according to

$$H = n \cdot (u \cdot \bar{h} - \bar{u} \cdot h). \tag{6.11}$$

Finally, the aberration or defect of distortion can be controlled via the fifth Seidel sum S_V. The maximum deviation of image point coordinates from the desired position D_{max} is given by

$$D_{\text{max}} = \frac{S_V}{2 \cdot n \cdot u}. \tag{6.12}$$

This value is equivalent to the absolute value as determined by equation (4.17), see section 4.1.6.

It turns out that applying equations (6.5)–(6.12) allows one to set up merit functions that consider any Seidel aberration. After solving the equations for the Seidel sum and defining the maximum acceptable aberration by the absolute value of the particular aberration, a target value for the appropriate Seidel sum is found. This value is the target defect d_t. An example for this procedure is shown in by the flow chart in figure 6.8.

---

[17] The aperture angle and the ray entrance height of the marginal ray are given by the non-overbarred formula symbols whereas overbarred formula symbols apply to the chief ray.

6.2.2 Direct optimisation via manual modifications

The use of numerical models and algorithms is not the only way for direct optimisation. Some reasonable basic considerations and manual modifications of an optical component, an optical system, or an opto-mechanical setup may also give good or sufficient results. Even slight modifications allow notable enhancements in some cases. Obviously, knowledge and understanding of the underlying mechanisms for the formation of optical aberrations becomes crucial for such modification. This is illustrated by the following examples:

Example 1
A simple reduction of the aperture stop diameter might result in a moderate decrease in illumination, but reduce spherical aberration significantly by masking the outer light rays of a ray bundle. Especially for strongly curved lens surfaces, this method is an adequate solution in many cases.

> FOCUS ON:
> Adjusting the stop diameter
> → Exercises **E15** and **E28**

Example 2
Shifting components is another manual approach for improving the image quality of an optical system. As explained in more detail in section 3.1.3, the variation of the aperture stop position is a good approach for the minimisation of coma by approximating the natural stop position. Moreover, a slight shift of a detector towards or off the lens or system can be sufficient to convert a non-diffraction-limited system to a diffraction-limited one.

> FOCUS ON:
> Optimisation via defocussing
> → Exercise **E26**

Example 3
Changing the material of a lens may also result in an improvement of the imaging quality. Especially for addressing chromatic aberration, this approach is well-established, but it can also be applied for monochromatic imaging tasks such as, for example, laser optics: the higher the index of refraction, the shorter the focal length of a lens with a given radii of curvature. And the shorter the focal length, the smaller the focal point diameter. This point is of specific interest in laser materials processing applications.

> **FOCUS ON:**
> Changing the lens material
> → Exercises **E14**, **E22** and **E 27**

6.2.3 Indirect optimisation

For **indirect optimisation**, the physical parameters of an optical component or system (e.g. radii of curvature, lens thickness or material, lens distances or air gaps) are not modified. Here, auxiliary correction optics are used instead. The most familiar example for such indirect optimisation via auxiliary correction optics is the use of spectacles or eyeglasses and contact lenses[18]. Here, either short-sightedness or long-sightedness[19] is corrected by shifting the eye's natural focal point position towards the image plane, i.e. the retina surface. This corresponds to a simple positive or negative defocus.

Not only the human eye but also technical optical systems may be corrected by auxiliary optics[20]. For instance, immersion oils or liquids are applied in order to increase the field of view or the aperture angle u of microscope lenses. The field of view can also be expressed by the numerical aperture NA according to

$$NA = n_s \cdot \sin(u) \tag{6.13}$$

where n_s is the index of refraction of the surrounding medium. By the use of an immersion liquid with an index of refraction of 1.5 in-between the front lens of the microscope optics and the object, the field of view is thus increased by a factor of 1.5. Another commonly used approach is the addition of an aplanatic meniscus (see section 3.7) to an existing optical system where the goal is to increase the numerical aperture without introducing additional aberrations[21].

6.3 Determination of manufacturing tolerances

The result of optical system design as described so far is the best, but idealised solution. This means that after evaluation, re-design and optimisation the system

---

[18] Defects of vision, a.k.a. ametropia, can also be directly corrected via laser-assisted *in situ* keratomileusis (LASIK) where a part of the cornea is removed via laser irradiation. The surface of the eye lens is thus directly re-shaped.

[19] Short-sightedness is also referred to as myopia whereas long-sightedness is also known as hyperopia.

[20] One famous example for the correction of a technical optical system is the Hubble Space Telescope. It was launched in 1990 but due to a manufacturing error, the primary mirror of this telescope featured an incorrect radius of curvature and, as a result, severe spherical aberration. This error was corrected in 1993 by installing an auxiliary optical system, the Corrective Optics Space Telescope Axial Replacement (COSTAR)—Hubble Space Telescope's spectacles or eyeglasses.

[21] The application of immersion liquid or oil to the gap between the front lens of a microscope and—in most cases—the plane object that is usually placed on a glass slide results in the formation of an additional plano-concave lens within the imaging optical path. Even though the field of view is increased in that vein, additional aberrations may thus occur due to the use of immersion liquids or oils.

represents a single, individual answer to a given imaging task. It is calculated on the assumption of fix or constant radii of curvature, thicknesses, indices of refraction etc. But, what if for a certain lot size, these essential parameters vary? In this case, quite a number of actual solutions appear as exemplified by Case example 6.3.

Case example 6.3: Impact of manufacturing tolerances

Task
A lot of 100 symmetric biconvex lenses with a radius of curvature of 100 mm and a thickness of 5 mm are manufactured where the lens material's index of refraction is 1.51. For this parameter set, the target effective focal length accounts for EFL = 98.87 ± 0.2 mm. As a start, we now assume the following acceptable deviations
  • ± 0.1 mm for the first radius of curvature,
  • ± 0.2 mm for the centre thickness, and
  • ± 0.002 for the index of refraction.

Are the listed tolerances well-chosen or do we have to reduce the tolerance range?

Solution
For the given acceptable deviations, the following focal length ranges can be calculated or simulated:
  • 98.82–98.92 mm for the variation of the radius of curvature of the first lens surface,
  • 98.84–98.91 mm for the variation in centre thickness, and
  • 98.49–99.16 mm for the variation in index of refraction.

First, it turns out that the maximum deviation in effective focal length is exceeded for each single deviation or tolerance class[22]. Second, it can be seen that a finite number of effective focal lengths occurs for the given tolerance ranges. We thus get a finite number of ideal solutions instead of a single one as obtained by optical system design without any consideration of manufacturing tolerances. Moreover, this high number of ideal solutions is extended by a finite number of non-ideal solutions that result for additional local influences such as surface defects (humps, scratches, pits, etc), glass defects (bubbles, striae, and so on) etc. Finally, position tolerances were not considered in the present example.

In reality and even for the tightest manufacturing tolerances, a batch of optical components does not feature a common effective focal length, aberration, or imaging quality. The analytical determination of these parameters is virtually impossible due to the high number of variables. The specification of the required

---

[22] In practice, all tolerances apply at the same time, so the resulting deviation may even be higher. In contrast, deviations could be compensated.

accuracy and manufacturing tolerances, also referred to as tolerancing, is thus a process that consists of several—and usually iterative—steps:

- **Sensitivity analysis**, i.e. the identification of critical surfaces of an optical component or system with a high impact on the imaging quality. As presented in sections 5.2 and 5.1.8, this analysis can be performed based on Seidel aberration tables or Seidel bar charts.
- Eventual re-design of the optical component or system by changing or adapting the radii of curvature, lens material[23] etc in order to avoid or reduce critical surfaces as identified and monitored in the course of a sensitivity analysis.
- **Monte Carlo simulation**, i.e. the analysis of the impact of variations in glass quality, surface accuracy, position etc (so-called defects) on the imaging quality including successive limiting of such defects until the target imaging quality is obtained.

The last step, Monte Carlo simulation, is a repetition of random sampling via appropriate computational algorithms. This computer-based approach is applied in order to obtain numerical results for mathematical tasks with a large number of input variables. These include all material defects, surface deviations and position errors as introduced in chapter 7, for example, variations in index of refraction and V-number, surface shape, or centring. In the course of Monte Carlo simulation, different tolerance classes including the corresponding tolerance values (i.e. the upper and lower limits of each defect) can be applied as listed in table 6.2.

The result of the application of these tolerance ranges during Monte Carlo simulation is a set of solutions that is calculated for each possible optical system defined by the tolerance values. One can easily imagine that for a complex system

Table 6.2. Examples for different tolerance classes applied for the evaluation of the impact of manufacturing tolerances on optical systems.

Tolerance class	Parameter (and particular tolerance range)			
	Radius of curvature in fringes	Centre thickness in microns	Index of refraction	Abbe number
Standard	±10	±250	±0.001	±0.5
Precision	±5	±100	±0.0005	±0.2
Extra precision	±1	±50	±0.0002	±0.1

---

[23] For the adaption of lens materials, quite helpful tools such as active glass maps [7] are in hand.

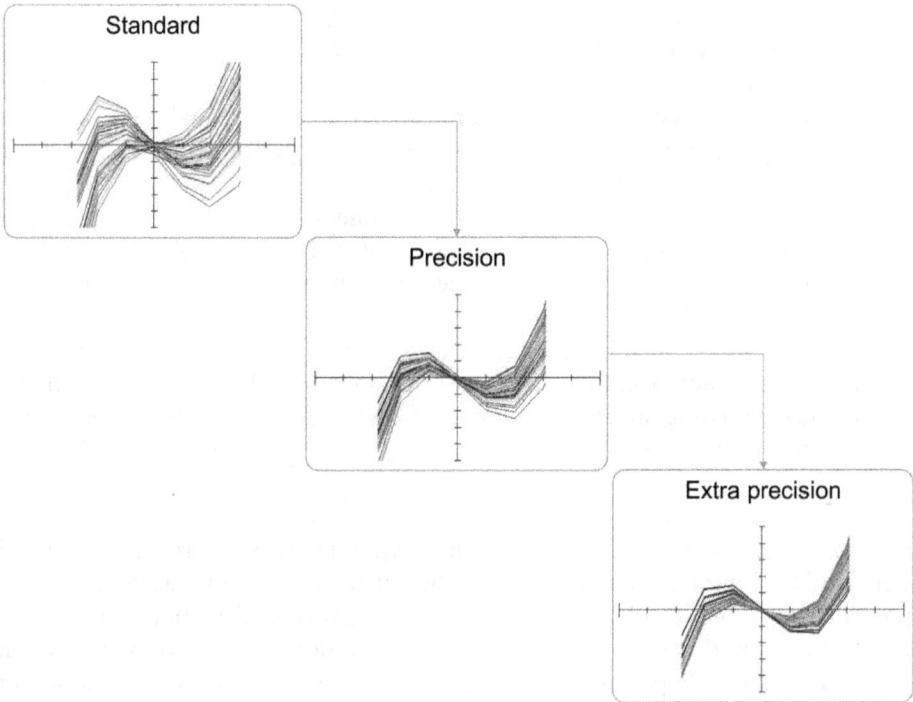

Figure 6.9. Impact of tolerance classes—standard, precision and extra precision—on the imaging performance as monitored by the transverse ray aberration diagram.

with, for example, 6 lenses and 12 surfaces, respectively, an extensive number of calculation steps and corresponding solutions occur.

The evaluation of the impact of manufacturing tolerances on imaging quality and performance can be carried out by the analysis of different graphs as presented in chapter 5. As shown in figure 6.9, not a single curve, but a curve cloud results since each possible solution within the defined tolerance range is shown.

In the case of insufficient performance as expressed by a strong scattering of the curves, the acceptable tolerance values are adapted until the calculated solutions are within an acceptable range[24]. The final tolerance values are then documented in the result of optical system design and tolerancing—a manufacturing drawing. An example of such a drawing is shown in figure 6.10. The meaning and specification of the elements and abbreviations shown here are introduced and explained in detail in chapter 7.

---

[24] Here, a certain knowledge of glass and optics manufacturing is crucial. A theoretically good solution might require tolerance values that cannot be realised in practice. Depending on the lens geometry and material, there are certain limits of feasibility and producibility.

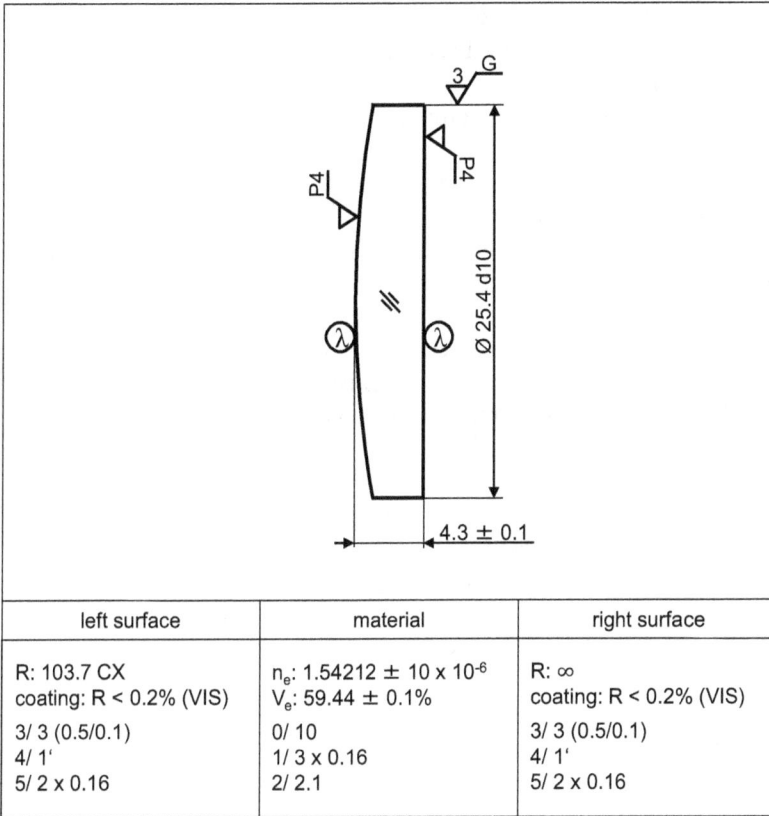

left surface	material	right surface
R: 103.7 CX coating: R < 0.2% (VIS) 3/ 3 (0.5/0.1) 4/ 1' 5/ 2 x 0.16	n_e: 1.54212 ± 10 x 10^{-6} V_e: 59.44 ± 0.1% 0/ 10 1/ 3 x 0.16 2/ 2.1	R: ∞ coating: R < 0.2% (VIS) 3/ 3 (0.5/0.1) 4/ 1' 5/ 2 x 0.16

Figure 6.10. Example of a manufacturing drawing of a single lens including all relevant tolerances required for production.

6.4 From task to solution—bottom line

- The very first step of optical system design is the analysis of a given imaging task and the identification of given and desired parameters.
- Simple calculations and considerations allow the determination of the focal length of a required optical system.
- A start system is an optical setup that promises the best imaging performance for a given imaging task.
- Start systems can be identified based on the field of view and the f-number of an imaging system.
- Based on an appropriate start system, an optical setup is evaluated and optimised in the course of an iterative process where also manufacturing tolerances may be considered.
- An optical system can be optimised directly or indirectly.
- Direct optimisation describes the modification of physical parameters such as radii of curvature or materials.

- Direct optimisation is based on the definition and application of a so-called merit function, taking variables (parameters to be changed) and defects (e.g. aberrations) into account.
- A merit function combines actual defects, target defects and defect tolerances; its result is a unitless value.
- The lower the value of the merit function, the better the imaging performance of an optical component or setup.
- An improvement in imaging quality via direct optimisation can also be realised by changing mechanical parameters such as the stop diameter or position.
- Indirect optimisation is carried out by adding auxiliary correction optics as, for example, aplanatic lenses to an existing system.
- The determination of manufacturing tolerances is the analysis of the impact of material or manufacturing errors on the imaging quality of an optical system.
- High-precision manufacturing tolerances may become crucial in order to preserve the required imaging quality.
- As a result, necessary manufacturing tolerances are identified, defined and documented.
- The final 'product' of lens design or optical system design is a manufacturing drawing.

References

[1] Gerhard C 2017 *Optics Manufacturing: Components and Systems* 1st edn (Boca Raton, FL: CRC Taylor & Francis)
[2] Smith W J 1992 *Modern Lens Design* 1st edn (New York: McGraw Hill)
[3] Schuhmann R G and Adams G 2001 *Proc. SPIE* **4441** 30
[4] Schuhmann R G and Adams G 2001 *Proc. SPIE* **4441** 37
[5] Kidger M J and Wynne C G 1967 *Opt. Acta* **14** 279–88
[6] Sturlesi D and O'Shea D C 1989 *Proc. SPIE* **1168** 90–105
[7] Schuhmann R G and Adams G 2004 *Proc. SPIE* **5249** 364

Chapter 7

Impact of manufacturing tolerances on imaging performance

The final product of optical system design is a manufacturing drawing including all relevant tolerances and allowances. Such tolerances are determined in the course of optics design as described in section 6.3. Hence, basic knowledge about glass and optics manufacturing is advantageous or even crucial. Optical components are mainly specified according to ISO 10110[1]. This standard includes information on material imperfections, surface form tolerances, centring tolerances, surface imperfection tolerances etc, where each defect or tolerance is indicated by its proper code number. This chapter gives an overview on relevant tolerances in glass and optics manufacturing including the particular impact on imaging performance and quality. Bulk defects in optical glasses as well as form deviations, surface defects and position tolerances of optical components are discussed. Moreover, the particular nomenclature according to the pertinent standard is explained.

7.1 Bulk defects in optical glasses

Quite a number of different glass bulk defects such as variations in index of refraction and V-number, stress birefringence, bubbles and inclusions, large-scale inhomogeneity in index of refraction, and short-range striae can arise in the course of the glass manufacturing process [1]. Such glass defects may cause blur, haziness and distortion of images as well as the formation of secondary ghost images. Moreover, scattering and absorption at smallest bubbles or impurities[2] contributes to a decrease in contrast transfer and imaging quality as well as a reduction in laser-induced damage threshold (LIDT) of laser optics. It is thus important to consider the impact of physical defects and the corresponding manufacturing tolerances for all-encompassing optical system design.

---

[1] ISO 10110 = 'Optics and photonics—Preparation of drawings for optical elements and systems'.
[2] It should also be noted that glass bulk defects might reach the surface during the precision optics manufacturing process and thus become surface defects.

7.1.1 Stress birefringence

Stress birefringence is indicated by the code number 0. It specifies the acceptable birefringence of a glass component that is induced by mechanical stress [2] resulting from the cooling procedure during glass making. The full nomenclature of birefringence is '0/A', where A is the acceptable maximum difference in optical path length, which arises due to birefringence. This difference is given in nanometres per 10 mm optical path length. The effect of stress birefringence means that an optical medium features two different indices of refraction. Unpolarised light is thus split into two different fractions with different light paths. Apart from the deformation of a transmitted wave front, this defect thus leads to the formation of two images, one for each polarisation direction of light[3], and an accompanying blurring of the image as shown in figure 7.1.

7.1.2 Bubbles and inclusions

The code number 1 indicates **bubbles** and **inclusions**, i.e. air or gas bubbles and solid inclusions[4] that are embedded in a glass component. Here, the total cross-sectional area of all bubbles and inclusions within a glass volume of 100 cm^3 is specified. The value is given in mm^2. The full nomenclature is $1/N \times A$, where N is maximum number of bubbles and inclusions and A is the acceptable cross-sectional area of each bubble or inclusion. In terms of imaging performance small bubbles within an imaging component can be described as micro ball lenses that cause refraction, deflection and scattering of incident light. Moreover, due to the size of bubbles in the range of some tens to some hundreds of microns, diffraction and interference can occur within the bulk material.

Figure 7.1. Example for birefringence of a calcite crystal.

[3] A well-known example for the formation of two images due to birefringence is the calcite crystal, see figure 7.1. One has to notice that here, the effect is due to intrinsic birefringence of the crystal material and the orientation of the crystal axes and not caused by mechanical stress.
[4] Inclusions are, for example, crystals or crystallites, non-molten residues from the used raw materials, debris from the furnace etc.

In contrast, inclusions normally are non-transparent particles, leading to absorption and scattering of light. All the mentioned effects cause a blurring of the image.

7.1.3 Inhomogeneity and striae

Deviations in index of refraction caused by **inhomogeneity** and **striae** are indicated by code number 2. Per definition, inhomogeneity is a large-scale deviation in index of refraction and thus refers to the entire volume of a component made of glass. In contrast, striae are local short-range deviations in index of refraction. Both types of deviations are described by a classification into classes. There are six classes (0–5) for homogeneity where the highest homogeneity is represented by class 6. Striae are classified into five classes (1–5) based on both the striae density—i.e. the share of striae in total optically active surface area of an optical component given in %—and the resulting wave front deformation. Class 5 indicates the highest purity. The full nomenclature of homogeneity and striae is '2/A × B', where A is the inhomogeneity class and B represents the striae class.

The consequence of inhomogeneity in index of refraction or striae in terms of imaging quality is a deformation of wave fronts as shown by Case example 7.1. Apart from such wave front deformation, inhomogeneity in index of refraction or striae represent a variation in index of refraction and V-number, respectively. This discrepancy may lead to notable differences between the theoretical performance as calculated based on the nominal values and the actual imaging quality.

Case example 7.1: Wave front distortion due to inhomogeneity

Task

A plane plate with a thickness of $t = 5$ mm is added to an existing optical setup[5] where the maximum acceptable increase in wave front distortion is 200 nm. We now have to determine the required homogeneity class of the glass used for this plane plate. The definition of the homogeneity classes is shown in table 7.1.

Table 7.1. Homogeneity classes including the corresponding acceptable deviation in index of refraction.

Homogeneity class	Acceptable deviation in index of refraction
0	$\pm 50 \times 10^{-6}$
1	$\pm 40 \times 10^{-6}$
2	$\pm 10 \times 10^{-6}$
3	$\pm 4 \times 10^{-6}$
4	$\pm 2 \times 10^{-6}$
5	$\pm 1 \times 10^{-6}$

---

[5] For instance, plane plates can be added to existing setups in order to correct present defocus by shifting a converging bundle of light towards the Gaussian image plane due to the effect of beam displacement.

Solution

Based on a known or given wave front distortion Δw, the underlying or causative deviation in index of refraction Δn can be calculated from

$$\Delta n = \frac{\Delta w}{2 \cdot t}. \tag{7.1}$$

It thus amounts to

$$\Delta n = \frac{200 \text{ nm}}{2 \times 5 \times 10^6 \text{ nm}} = 20 \times 10^{-6}$$

in the present case. This deviation can also be expressed as $\pm 10 \times 10^{-6}$. The plane plate should thus be made of a glass with the homogeneity class 2, or higher.

7.2 Form deviations and surface defects of optical components

In optical system design optically active surfaces are initially assumed to be perfectly shaped. This means that a spherical surface is a perfect segment of a sphere and a plane surface is perfectly flat. In practice, a real surface features different surface defects such as contour inaccuracies, surface damages (e.g. scratches and digs) and a certain residual surface roughness[6]. Such defects inevitably occur in the course of the manufacturing process of optical components.

7.2.1 Surface accuracy

Surface accuracy describes the deviation of a surface shape with respect to the target surface shape (i.e. the theoretical reference profile, e.g. the surface of a sphere for spherical lens surfaces). This defect is indicated by code number 3 where three parameters—*A, B* and *C*—are defined [1]:

    A. The amount of deviation from the target surface. In the case of a spherical lens surface, the value of the difference in radii of curvature is specified. This difference is referred to as sagitta.

    B. The deviation from the target shape. For instance, a lens surface can feature two different radii of curvature perpendicular to each other. The surface is thus slightly elliptical or toric, but not spherical. The difference in both radii of curvature gives the deviation from the target shape.

    C. The fine contour error. This parameter covers the smallest local defects such as humps or pits. During measurement, such defects become visible by a local deformation of the measured interference pattern or wave front. The value of the fine contour error is given by the ratio of the local deformation width of a Newton fringe[7] and the general distance between neighboured Newton fringes.

---

[6] In addition to surface roughness, an error of higher order, the so-called surface waviness is existent.

[7] A Newton fringe—named after the English mathematician and physicist *Sir Isaac Newton* (1642–1726)—is the basic element of an interferometric pattern.

Surface accuracy is determined within a pre-defined test area[8]; its full nomenclature is $3/A(B/C)$. As listed above, A is the maximum acceptable sagitta given in interference fringes for a defined test wavelength, B is the maximum acceptable deviation from the basic target shape with respect to A, and C is the fine contour error.

Contour inaccuracies are thus a deviation of the actual radius of curvature from the calculated one. This might consequently cause a deviation of the actual focal length from the calculated one. Since for high values for B toric or elliptical lens surfaces occur, even wave front distortion or astigmatism (see figure 7.2) can be formed without any inclined incidence of light.

> **FOCUS ON:**
> Toric lens surfaces
> → Exercise **E30**

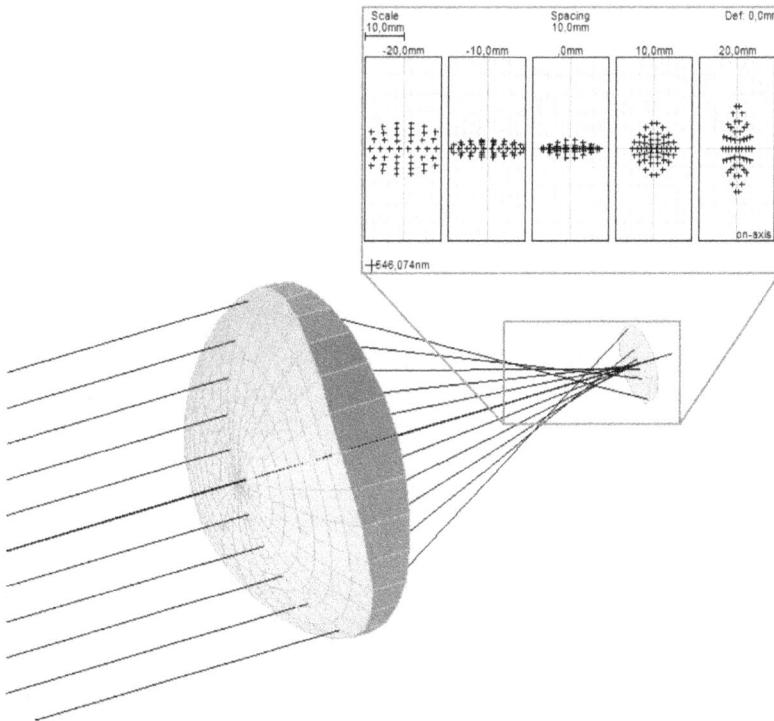

Figure 7.2. Visualisation of the formation of astigmatism caused by a toric lens surface.

---

[8] If not otherwise specified, the test area is normally given by a concentric circle that includes 80% of the total optically active surface area. The lens border, where it is finally mounted, is thus excluded from testing.

7.2.2 Surface cleanliness

Surface cleanliness addresses the cleanliness of polished glass surfaces taking surface defects and impurities into account. It is characterised by code number 5 where both the acceptable number and size of defects such as scratches, digs and stains on optically active surfaces are quantified. This is specified by the full nomenclature, $5/A \times B$, where A is the maximum acceptable number of defects and the parameter B quantifies the area of each single defect. Apart from ISO 10110, the U.S. standard MIL-PRF-13830B is commonly used for specifying surface cleanliness [3]. Here, the maximum acceptable length or dimension, but not the number of surface defects is specified. Since the area of surface defects depends on the test area, this standard is a relative description and not an absolute one, such as ISO 10110. Moreover, the MIL-PRF-13830B distinguishes scratches from digs. The nomenclature is $A-B$ scratches-digs. The parameter A is not definitively specified, but usually represents the width of scratches given in microns [4]. Further, the number of scratches is not clearly specified, and the maximum length of scratches can amount to a quarter of the test area diameter. The parameter B represents the maximum acceptable diameter of digs given in hundredths of a millimetre. The maximum number of digs, however, is not specified. Thus, this standard is quite controversial [5, 6].

Regardless of the approach for specification, surface defects such as damage or pollution have a notable impact on imaging quality. Surface-adherent non-transparent pollutants may absorb or—for smaller contaminants—scatter incoming light, and scratches can cause multiple reflections and further scattering. Absorption at scratches, digs and stains may further cause a decrease in LIDT of laser optics.

7.2.3 Surface roughness

Any polished surface features a certain residual roughness. There is no code number for describing roughness, but different classes of surface roughness are specified for both polished (i.e. optically active) surfaces and ground surfaces[9]. According to ISO 10110, the roughness of a polished surface is indicated by the expression PX where P means 'polished' and X is the grade of polishing, ranging from P1 to P4. The residual roughness of a P1-polished surface is some hundreds of nanometres whereas for the highest grade of polishing, surface roughness amounts to merely some nanometres. Moreover, so-called super polished surfaces with a roughness of a few angstrom[10] are producible nowadays.

Surface roughness is a type of optically active microstructure. Its impact on incoming light strongly depends on wavelength. The most disturbing effect is diffuse reflection and transmission, i.e. scattering of light, leading to blur, haziness and distortion of images as well as to the formation of ghost images.

---

[9] For example, the outer cylinder of a lens or the side faces of prisms.
[10] 1 angstrom = 1 Å = 100 pm.

7.3 Position tolerances

Up to now, we have discussed 'pure' optical defects. However, a lens or optical component is also described by its mechanical dimensions and geometry. For instance, the centre thickness has a certain influence on the effective focal length as shown by equation (3.4). Soft tolerances may thus lead to a noticeable variation in focal length within a batch size during production. Another relevant type of defect is position errors that are expressed or simulated by tilt and decentre of components in optical system design. Such tilt or offset of optics with respect to the system axis or other components as well as distance errors between optical components can arise in the course of manufacturing, mounting, or cementing. Since in the end, a simulated optical system has to be realised in the form of an opto-mechanical setup, the most common position tolerances including the particular impact are presented hereafter.

7.3.1 The centring error

Strictly speaking, the **centring error** is a form deviation of a lens since it describes the angular deviation between its optical and mechanical axis, i.e. the cylinder axis of the lens border cylinder. According to ISO 10110, this error is indicated by code number 4 where the acceptable angular deviation is particularised. The nomenclature is $4/X^Y$; X gives the value of angular deviation and Y its unit, either arc minutes (') or arc seconds ("').

Centring becomes an issue during mounting of a lens into a mount. The most common method is the so-called **stacking** method where lenses and distance rings are stacked into a tube and finally fixed by a threaded ring[11]. The lens thus aligns mechanically along the inner cylinder of the mount or tube—and consequently along the lens border cylinder. Let us now assume that the lens is decentred and features an angular deviation between the mechanical and the optical axis: as a result, the optical axis is tilted. As shown by the example in figure 7.3, the image spot geometry is distorted in such a case.

Another possible reason for a lens tilt is the non-circularity of bevels. Lenses are usually bevelled in order to realise a protection chamfer that is slanted by 45° with respect to the lens border cylinder surface. A non-circular bevel is not perfectly ring-shaped as it should be, but features different bevel leg lengths as visualised in figure 7.4.

Since the lens rests on the bevel, this error leads to a tilt α of the lens after fixing it in a mount or on a holder. This tilt depends on the maximum leg length l_{max} and the minimum leg length l_{min} as well as the radius of curvature R_c of the lens according to

$$\alpha = \frac{l_{max} - l_{min}}{2 \cdot R_c}. \tag{7.2}$$

---

[11] This threaded ring is the outermost and thus visible mechanical component of a stacked optical system. It is usually a thin blackened ring with two slits or notches for screwing. It can easily be seen on binoculars or camera lenses.

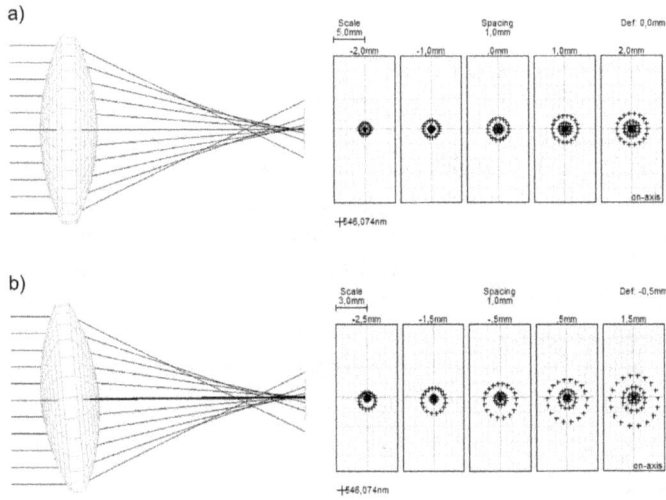

Figure 7.3. Visualisation of the centring error by the comparison of a (a) perfectly aligned lens and a (b) decentred one including the particular corresponding spot diagram.

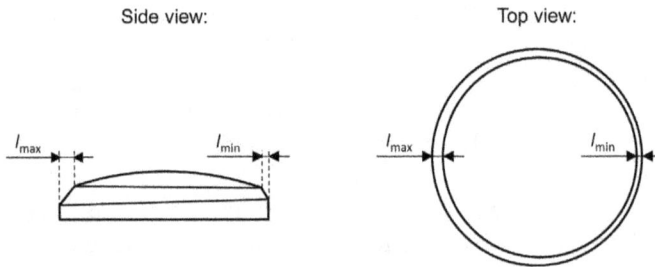

Figure 7.4. Visualisation of non-circular lens bevels with a maximum and minimum leg length l_{max} and l_{min}, respectively.

Case example 7.2: Tilt due to non–circular chamfers

Task
In an optical setup, the maximum tilt of a bevelled plano-convex lens with a radius of curvature of 45 mm is 17 arc minutes. Determine the maximum acceptable non-circularity $\Delta c = l_{max} - l_{min}$ of the bevel.

Solution
The desired non-circularity can be calculated after solving equation (7.2) for the maximum acceptable non-circularity:

$$\Delta c = l_{max} - l_{min} = \alpha \cdot 2 \cdot R_c = 0.005 \cdot 2 \cdot 45 \text{ mm} = 0.45 \text{ mm}.$$

Here, the tilt angle has to be inserted in radian where $17' \approx 0.29° \approx 0.005$ rad.

7.3.2 Defects of cemented lens groups

Cemented lens groups consist of at least two lenses that are contacted and cemented. The most familiar example is an achromatic lens. Here, the surfaces with the same absolute value of radius of curvature (a convex surface and a concave one) are brought in direct contact after applying a drop of so-called fine cement[12] to one surface. Both lenses are then pressed together, leading to the formation of a thin closed cement layer in between both lens surfaces. After alignment where the optical axes of both lenses are matched, the cement is finally cured via UV-irradiation, tempering, or storing, depending on the type of cement.

In the course of such cementing, different errors can occur:
- The optical axes of the involved lenses can be tilted.
- The optical axes of the lenses can be displaced laterally.
- The involved optical axes are both tilted and displaced laterally[13].

In any case, incident light that propagates parallel to the optical axis is deviated when passing the lens group. Consequently, the focal point is found off the optical axis and additionally deformed as shown in figure 7.5.

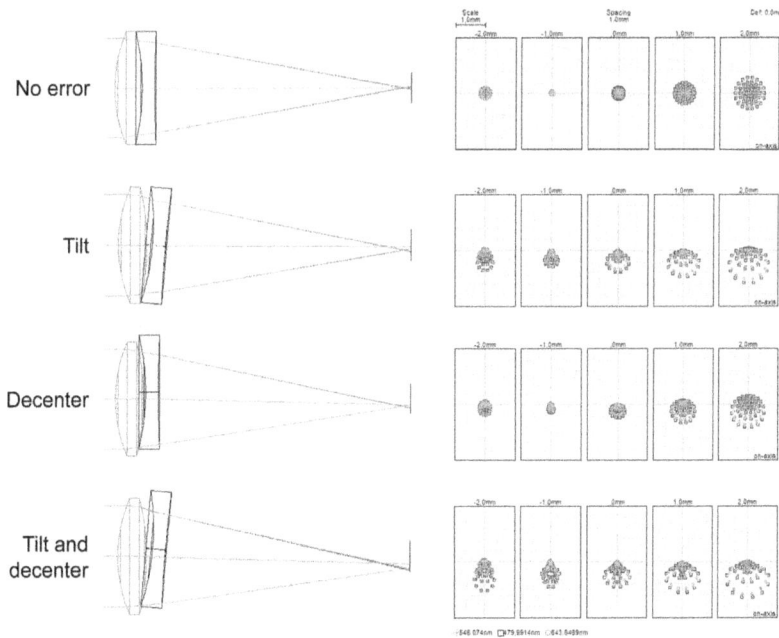

Figure 7.5. Visualisation of the impact of different cementing errors—tilt, decentre and the combination of both—on the spot diagram of a doublet.

---

[12] Optical fine cements are usually synthetic two-component polymers. In former times, the natural resin of North American firs was used—the so-called Canada balsam.
[13] In optics manufacturing, this error is known as cement wedge.

7.3.3 Defects of mounted opto-mechanic assemblies

As shown in section 7.3.2, position tolerances in opto-mechanical groups might have a significant impact on the performance and imaging quality. This does not only apply to the process of cementing, but also mounting. During the assembly of opto-mechanical systems, quite a number of position errors can arise due to manufacturing errors or inappropriate tolerances of mechanical components and elements:

- Inaccurately structured mounts or—in the case of stacked systems—insufficient tolerancing of spacer rings can result in distance errors. Here, the air gap between lenses and lens groups is either too thin or too thick[14]. This error may disturb the entire path or rays in an optical system as shown in figure 7.6.
- In staged mounts, the particular optical elements rest on bearing surfaces. A tilted or wedged bearing surface thus transfers its tilt to the lens, giving rise to a displacement and distortion of light bundles.
- Out-of-centre inner cylinders of mounts or bearing surfaces can cause a lateral offset of optical components. Such offset also results in deflection and deformation of light bundles.

FOCUS ON:

Mounting errors and tolerances

→ Exercises **E31 - E34**

Such possible defects of mechanical mounts and holders have to be considered by appropriate position tolerances[15]. It can finally be stated that ultimately, the imaging quality and performance of a physical optical system also depends on the manufacturing accuracy of the required mechanical components[16].

---

[14] Note that an air gap behaves like an optical component even though it is not an imaging element by definition. However, an air gap between two lenses has two curved interfaces and an index of refraction of its 'bulk material', air, where $n_{air} = 1.0003 \approx 1$.

[15] Mechanical tolerances are specified according to the Geometric Dimensioning and Tolerancing (GD&T) standard. It is defined by the US-Geometric Product Specification-Standard ASME Y 14.5 on the one hand and ISO 1101 on the other hand. These standards cover the form, profile, position, orientation and run-out of mechanical components.

[16] Position tolerances are not only of importance for housed opto-mechanical systems such as binoculars or microscope and camera lenses; in science, quite complex optical setups, partially consisting of several tens or even hundreds of different optical components are realised by fixing the components on mechanical holders manually. Here, decentre of some hundreds of microns and a tilt of a few degrees can easily occur. The alignment of such setups thus requires experience, skilfulness...and patience.

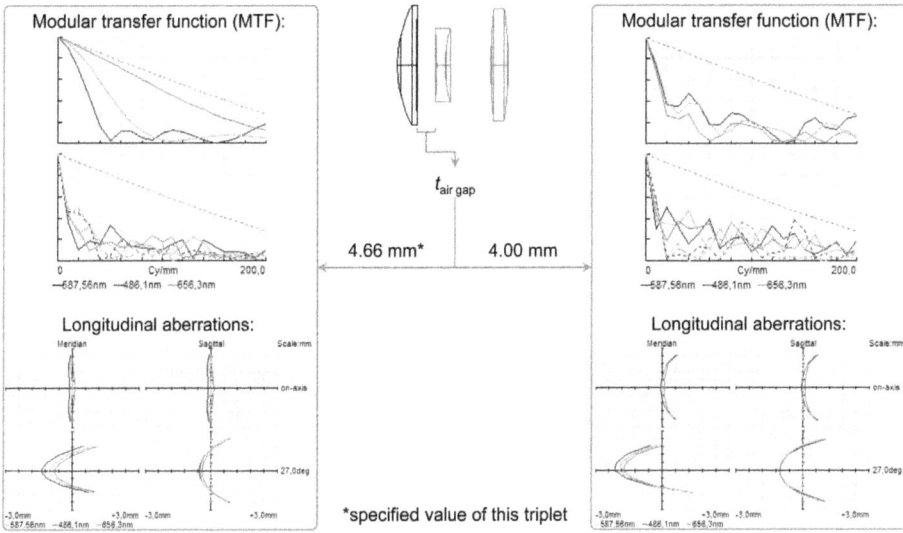

Figure 7.6. Visualisation of the influence of a variation in air gap thickness $t_{air\ gap}$ in a split triplet on the modular transfer function and longitudinal aberrations.

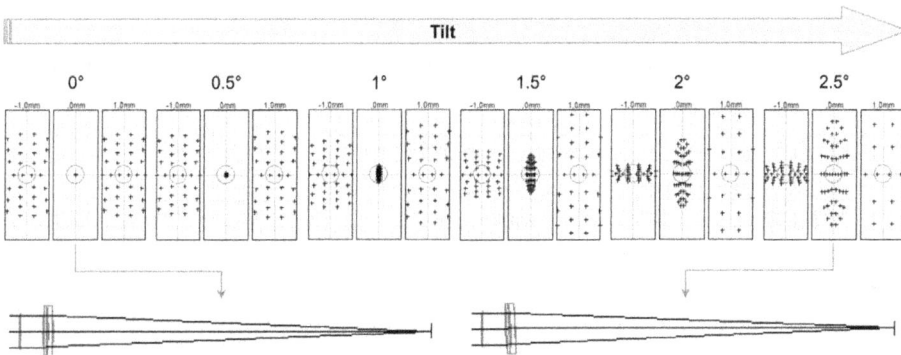

Figure 7.7. Determination of the maximum acceptable tilt angle of an optical element for diffraction-limited imaging onto the Gaussian image plane.

Case example 7.3: Tilt of optical components

Task

An achromatic doublet with an effective focal length of EFL = 500 mm is mounted in a tube. The goal is to realise diffraction-limited imaging of a collimated incoming light bundle with a diameter of 40 mm. What is the maximum acceptable tilt of the doublet?

Solution

A possible solution is shown in figure 7.7. Here, an achromatic doublet with an effective focal length of 500.39 mm and a diameter of 63 mm was chosen (catalogue number 322 230 000 in the WinLens lens database). In order to gain information on diffraction-limitation and to evaluate the impact of tilt on the size of the image spot in the Gaussian image space, the Airy disc is overlaid in the spot diagram.

According to this evaluation of the spot diagrams, the limit for tilt is thus found at a tilt angle of approximately 1°. For higher angles, the spot size exceeds the diameter of the Airy disc.

Comprehension question 7.1: Interpretation of spot diagrams

In the spot diagram shown in Figure 7.7, an interesting effect can be observed for tilt angles of 2° or higher. Describe and interpret this effect.

Answer: At a tilt angle of 2°, two oval image points occur, one about one millimetre in front of the Gaussian image plane and another one right in the Gaussian image plane. Both are rotated by 90° to each other. This observation indicates that astigmatism occurs in the case of higher tilt angles.

7.4 Impact of manufacturing tolerances on imaging performance—bottom line

- Manufacturing tolerances may significantly affect the imaging quality and performance of optical components and systems.
- The tolerances of optical components are mainly specified according to the standard ISO 10110.
- According to ISO 10110, the most important defects of an optical component are specified by the code numbers and corresponding parameters.
- There are different types of defects,
    - bulk defects of optical glasses,
    - form deviations and surface defects of optical components, and
    - positioning errors.
- Stress birefringence specifies the acceptable birefringence of a glass component that is induced by mechanical stress; it is indicated by code number 0.
- The number and size of acceptable air or gas bubbles and solid inclusions that are embedded in a glass component are specified by code number 1.
- Deviations in index of refraction caused by inhomogeneity and striae are indicated by code number 2.
- Surface accuracy—indicated by code number 3—describes the deviation of a surface shape with respect to the target surface shape.
- Surface cleanliness quantifies the acceptable number and size of surface defects such as stains or scratches. According to ISO 10110, it is indicated by code number 5.

- Surface cleanliness can also be specified according to U.S. standard MIL-PRF-13830B.
- The surface roughness class indicates the acceptable residual roughness of polished and ground glass surfaces.
- The centring error as indicated by code number 4 gives the deviation of the optical axis of a lens and the mechanical axis of its outer cylinder.
- Non-circular protection bevels lead to a centring error.
- During cementing of lens groups, several errors can occur, leading to a tilt of the optical axes of the involved single components.
- Apart from manufacturing tolerances in optics manufacturing, the tolerances of mechanical components of an opto-mechanical setup are of significant importance.

References

[1] Gerhard C 2017 *Optics Manufacturing: Components and Systems* 1st edn (Boca Raton, FL: CRC Taylor & Francis)
[2] Adams L H and Williamson E D 1919 *J. Wash. Acad. Sci.* **9** 609–23
[3] Aikens D M 2010 *Proc. SPIE* **7652** 765217
[4] Aikens D M 2010 OSA Technical Digest OTuA2
[5] Young M 1989 *Proc. SPIE* **1164** 185–90
[6] Young M 1983 *Proc. SPIE* **0406** 12–22

IOP Publishing

Lens Design Basics
Optical design problem-solving in theory and practice
Christoph Gerhard

Chapter 8

Hands-on training

Tell me and I forget.
Show me and I remember.
Let me do and I understand.
—*Confucius*

This chapter contains 34 exercises sorted by the topics 'Basic considerations and definition of start systems', 'Creating and evaluating lenses and systems', 'Optimisation of optical systems', and 'Simulation of manufacturing errors and tolerances'. It starts with an introduction of the free software used, covering its handling and most important functions. The solutions of the exercises are provided in the appendix of this book.

8.1 Introduction to used software

For the hands-on training via the exercises given in this chapter, two different software tools are used: the layout tool **PreDesigner**[1] and the ray tracing program WinLens Basic[2] (hereafter simply referred to as '**WinLens**') [1–3]. The software as well as the Lens Library—a collection of different optical systems—can be downloaded free of charge, even without any registration. The install and update files are provided by Qioptiq, an Excelitas Technologies company, and are currently found in the company's online shop[3]. Both programs are quite intuitive and self-explanatory; the handling of

---

[1] PreDesigner is also available as an Android App, the 'Lens Calculator' that can be found in the Google Play Store.

[2] WinLens Basic is the free ray tracing program of the software package WinLens™3D. Its first version was released in 1993.

[3] At the time of printing of this book, the corresponding link was www.qioptiq-shop.com/Optik-Software/Winlens-Optical-Design-Software. However, since the link may change in the course of time, the software can alternatively be found easily by searching for the terms 'winlens', 'predesigner' and 'download' using any search engine.

the software can thus be exercised via learning by doing. However, the very first steps and the basic functions are explained in the following two sections. Moreover, additional information on the software can be found on the website of the programmer, the Optical Software Company, www.opticalsoftware.net. Here, quite a number of commented screencasts, videos, tutorials and articles explaining the basic functions of PreDesigner and WinLens as well as the set up of complex systems are provided.

8.1.1 The calculation tool PreDesigner

PreDesigner is a calculation tool that is intended to help solve those 'what if' questions at the start of a lens design, without the need for a real design. It facilitates basic considerations for the determination of conjugated parameters and relevant additional information. The principal mathematical model applied in PreDesigner is the paraxial imaging model, but it also includes the possibility to perform calculations of Gaussian beams.

When inserting three key parameters such as the object distance and size parameters[4], the required lens layout is displayed as a schematic drawing, see figure 8.1. Moreover, all related values are given in tabular form. This includes object and image distances, the focal length, track and magnification as well as object and image sizes and angles. After adding an aperture parameter, e.g. the

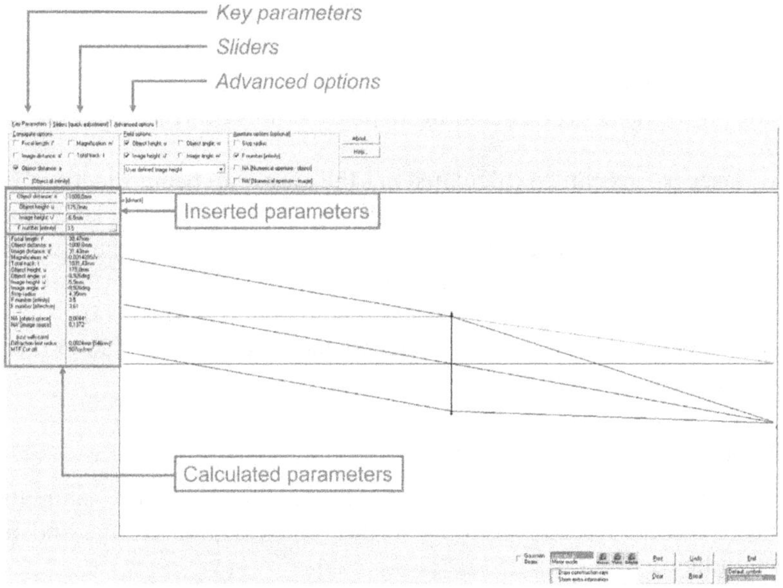

Figure 8.1. User interface of the calculation tool PreDesigner.

---

[4] Note that in PreDesigner, the object and image distance are abbreviated u and u', respectively (instead of y and y', respectively).

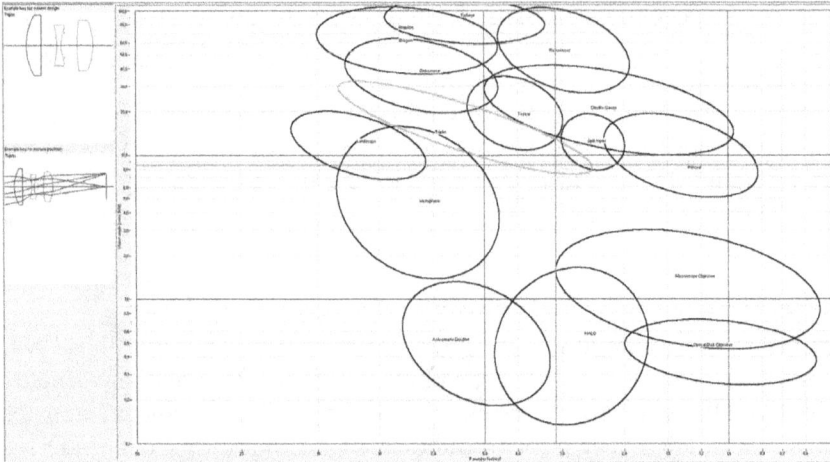

Figure 8.2. The f-number versus field plot provided in PreDesigner.

aperture stop radius or the f-number, the resulting Airy disc radius as well as the MTF cut-off frequency are further calculated. Sliders allow one to make quick changes to the key values where the layout and spreadsheet update in real time[5].

When the layout is ready, an f-number and a field angle will be part of the entire description of the optical setup. To help move on to an actual start design or system, PreDesigner shows an f-number versus field plot as introduced in section 6.1.2 where the typical performance region for different families of lenses is shown as a series of ellipses. The location of the current design is highlighted, thus suggesting an appropriate initial lens type including a preview of the lens setup as shown in figure 8.2.

Finally, laser beams can be calculated via the Gaussian beam function. Here, the particular classical parameters are replaced by relevant measures of laser beams. For instance, the object size is substituted by the beam waist radius of the raw beam. The calculated parameters include relevant characteristics of laser irradiation as, for example, the Rayleigh length[6]. As shown in figure 8.3, important background information such as the laser wavelength or the beam quality factor M^2 can be defined easily.

8.1.2 The ray tracing software WinLens

WinLens is an easy-to-use ray tracing software for the setup, simulation and evaluation of optical systems. Its start surface and essential editors and functions are shown in figure 8.4.

---

[5] This function clarifies the meaning of 'conjugated parameters'. When varying the object distance or height, the image distance or height adapt accordingly and vice versa. This function thus visualises the interconnection of parameters in the object and image space—just try for any system...

[6] The Rayleigh length z_R gives the distance from the laser beam waist to the position where the laser beam radius is $\sqrt{2}$-times the beam waist radius.

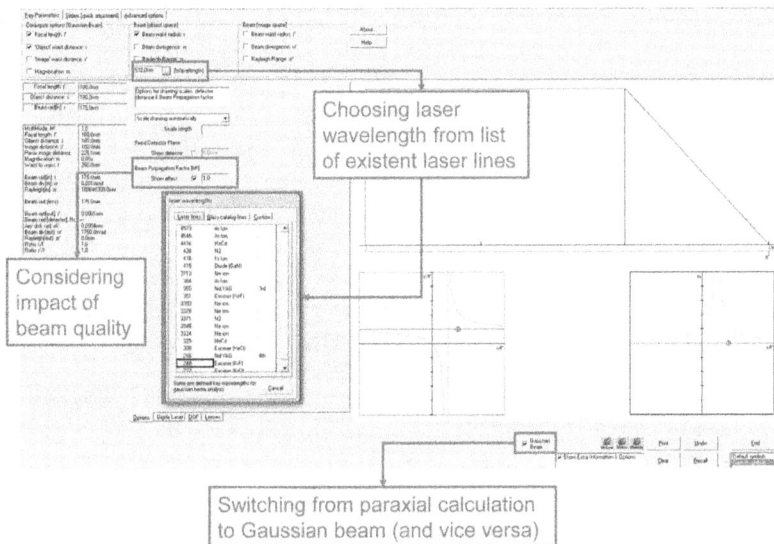

Figure 8.3. Overview of the Gaussian beam function in PreDesigner.

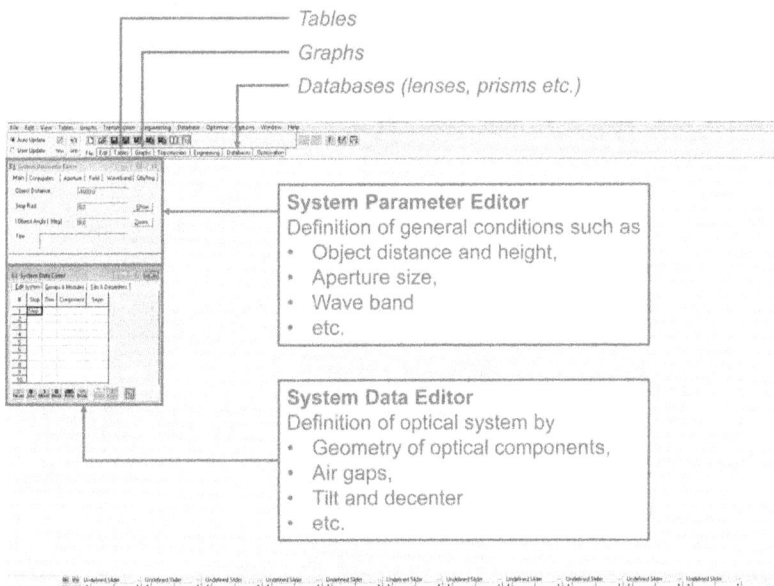

Figure 8.4. User interface of the simulation software WinLens Basic.

The given or conjugated parameters as, for example, determined using PreDesigner are entered in the System Parameter Editor. This editor contains six different index cards as listed in table 8.1 and shown in figure 8.5.

Optical components and systems are defined and set up in the System Data Editor. As listed in table 8.2 and shown in figure 8.6, this editor contains three index cards.

Table 8.1. Overview of the index cards in the WinLens System Parameter Editor including the particular function.

Name	Function
Main	Overview of defined system parameters
Conjugates	Definition of object or image distance, magnification, etc
Aperture	Definition of stop radius (0 per default), numerical aperture, f-number, etc
Field	Definition of object and image angle or size
Wave band	Definition of considered wave band (UV, vis, IR or user-defined)
Obj/Img (= object/ image space)	Definition of indices of refraction in object and image space, definition of curved detectors

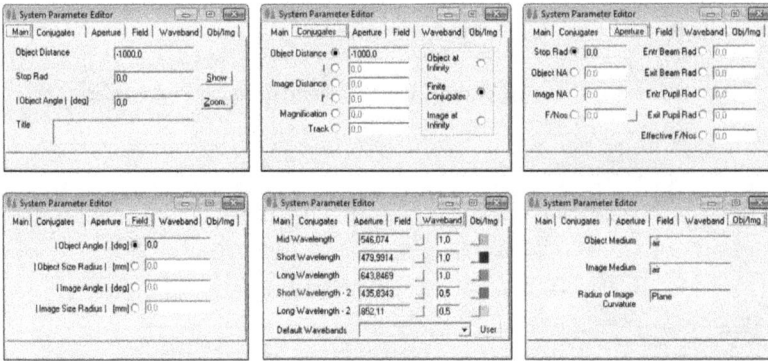

Figure 8.5. Overview of the index cards in the WinLens System Parameter Editor.

Table 8.2. Overview of the index cards in the WinLens System Data Editor including the particular function.

Name	Function
Edit system	Insertion of optical components including definition of orientation and distances
Groups and modules	Combination of several components to lens groups (e.g. achromatic doublets)
Tilts and decentres	Definition of tilts and decentres of optical components or single interfaces

The insertion of optical components or systems into the System Data Editor can be performed in different ways:
- using pre-defined single components,
- choosing catalogue optics from the included database,
- inserting the order number of a catalogue component, or
- using pre-defined optical systems from the Lens Library.

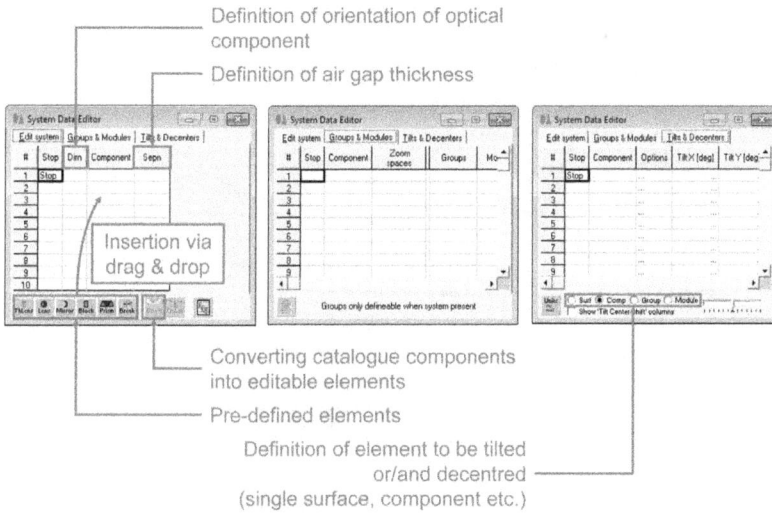

Definition of orientation of optical component

Definition of air gap thickness

Insertion via drag & drop

Converting catalogue components into editable elements

Pre-defined elements

Definition of element to be tilted or/and decentred (single surface, component etc.)

Figure 8.6. Overview of the index cards in the WinLens System Data Editor—edit system (left), groups and modules (middle), and tilts and decentres (right)—including the most important functions.

Pre-defined single components can be found at the bottom of the edit system index card of the System Data Editor. These components, i.e. thin lenses, thick lenses, mirrors, blocks, prisms and coordinate breaks are inserted into the component column via drag and drop. Then, a new window opens automatically where the particular component can be defined. As an example, figure 8.7 shows the definition of a lens by specifying its shape and dimension as well as its material. The material can be chosen from a catalogue of glass manufacturers. Here, one can search for the name of the glass, its index of refraction, its V-number and other related parameters as shown in figure 8.9. Once the lens is defined, a preview of its basic geometry and effective focal length is provided[7]. By clicking the OK button, the lens is finally displayed by the lens drawing, see figure 8.8.

Comprehension question 8.1: Lens drawing

Even though the lens is displayed in Figure 8.8, no light ray is traced or shown, neither in the lens drawing nor in any evaluation graph...why?

Answer: *Per default, the aperture stop radius is zero when opening WinLens, see Table 8.1. Hence, no rays can be traced; an aperture parameter has to be defined before.*

---

[7] This function is very helpful! It allows one to control if the proper algebraic signs were applied to the particular radii of curvature.

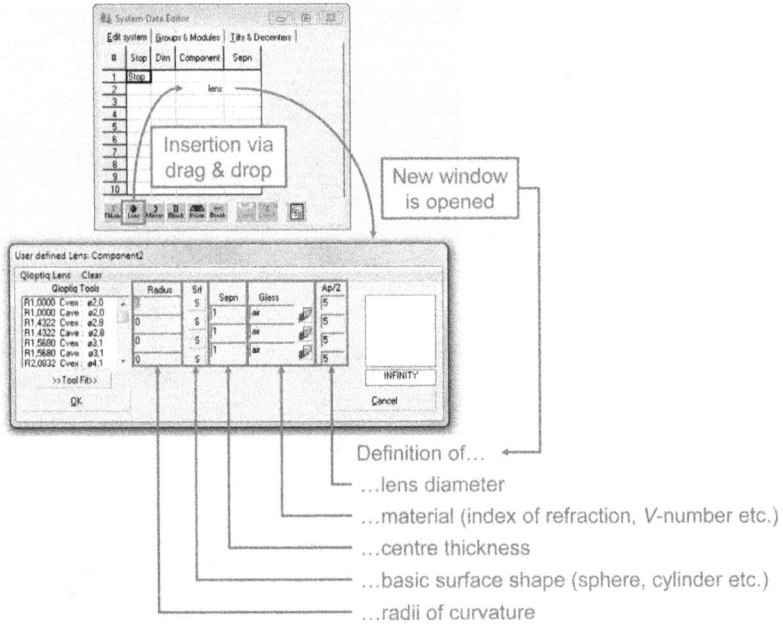

Figure 8.7. Definition of a lens by inserting a pre-defined element via drag and drop and entering relevant lens parameters.

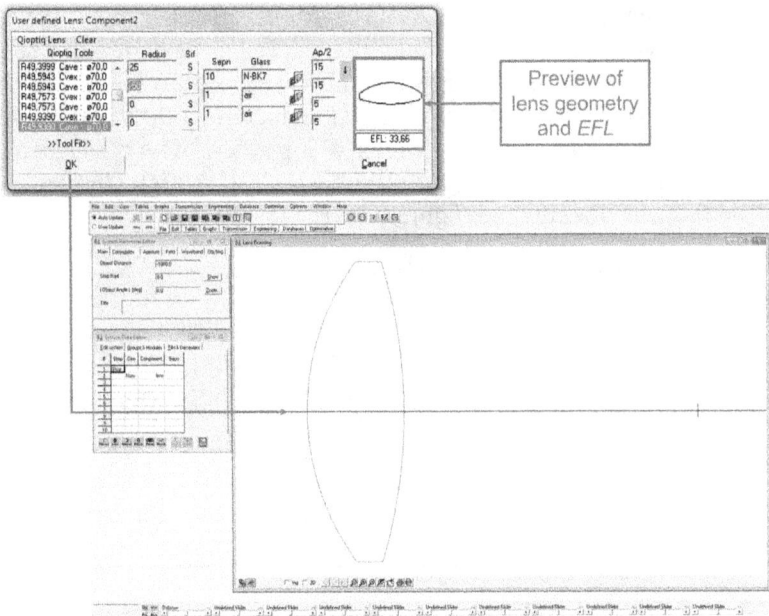

Figure 8.8. Example for a thick lens simulated based on a pre-defined element.

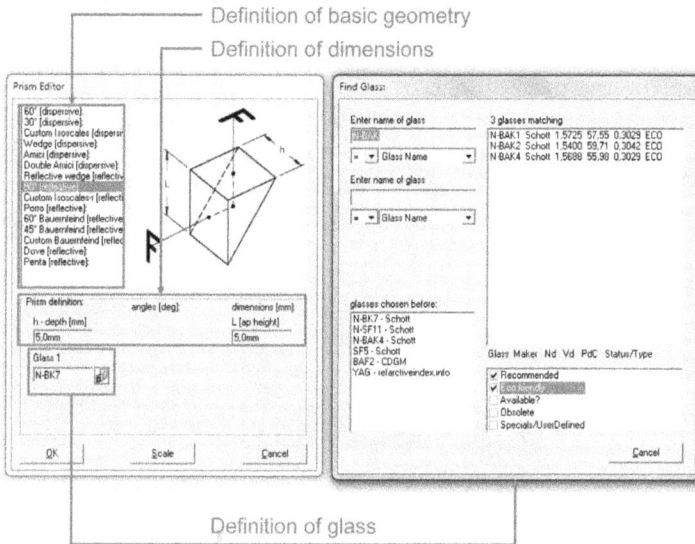

Figure 8.9. Definition and insertion of prisms using the Prism Editor.

Apart from lenses, a number of other pre-defined elements can be chosen. Figure 8.9 shows how to define a prism in the Prism Editor. The basic prism geometry is chosen from a list[8] and subsequently, the prism dimensions and its material are specified. The specification of the material is performed by a selection tool where glasses can be searched for by the glass name etc, as already mentioned above. One has to notice that after inserting a prism (or a tilted mirror), the optical axis is folded automatically and the subsequent optical elements are arranged along the new, folded axis.

Another possibility to insert optical components is the use of databases where a number of catalogue optics, i.e. lenses, prisms and diffraction gratings are provided. If the order number of the particular optical element is known, it can easily be added to the setup by entering the number into the component column of the System Data Editor. Moreover, complex optical systems such as triplets, Gauss lenses, telecentric lenses, tesar lenses etc, can be opened from the Lens Library.

The evaluation of the optical component or system can be performed with the aid of tables or graphs and diagrams as presented in detail in chapter 5 and shown in figure 8.10. One should notice that some diagrams (and functions) are disabled in the

---

[8] There are quite a few different prism types, categorised by function (dispersive prisms, reflective prisms, polarisation prisms etc) or geometry (penta prism, 90° prism etc). In some cases, the prism is named after its inventor, for example

- the Amici prism, named after the Italian astronomer *Giovanni Battista Amici* (1786–1863),
- the Porro prism, named after the Italian engineer *Ignazio Porro* (1801–1875), or
- the Bauernfeind prism, named after the German geodesist *Karl Maximilian von Bauernfeind* (1818–1894)

free WinLens Basic program. This includes, for example, the ghost image plot, the footprint plot (and the entire automatic optimisation function).

Apart from the essential functions—setting up and evaluating components and systems—WinLens offers quite a number of further tools and features. The global parameters that are accessed via a right mouse click allow for the definition of the number of traced and displayed rays, considered wavelengths and object points. In the drawing options, see figure 8.11, the paraxial items can be chosen, the displayed

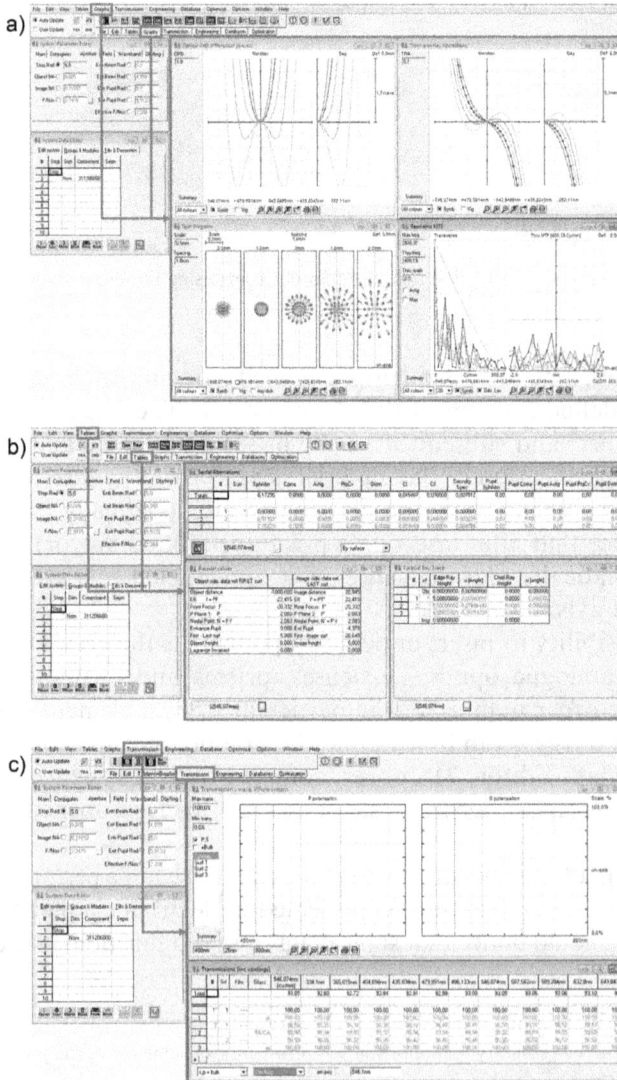

Figure 8.10. Examples of different evaluation tools in WinLens: (a) graphs and diagrams, (b) tables and (c) transmission data.

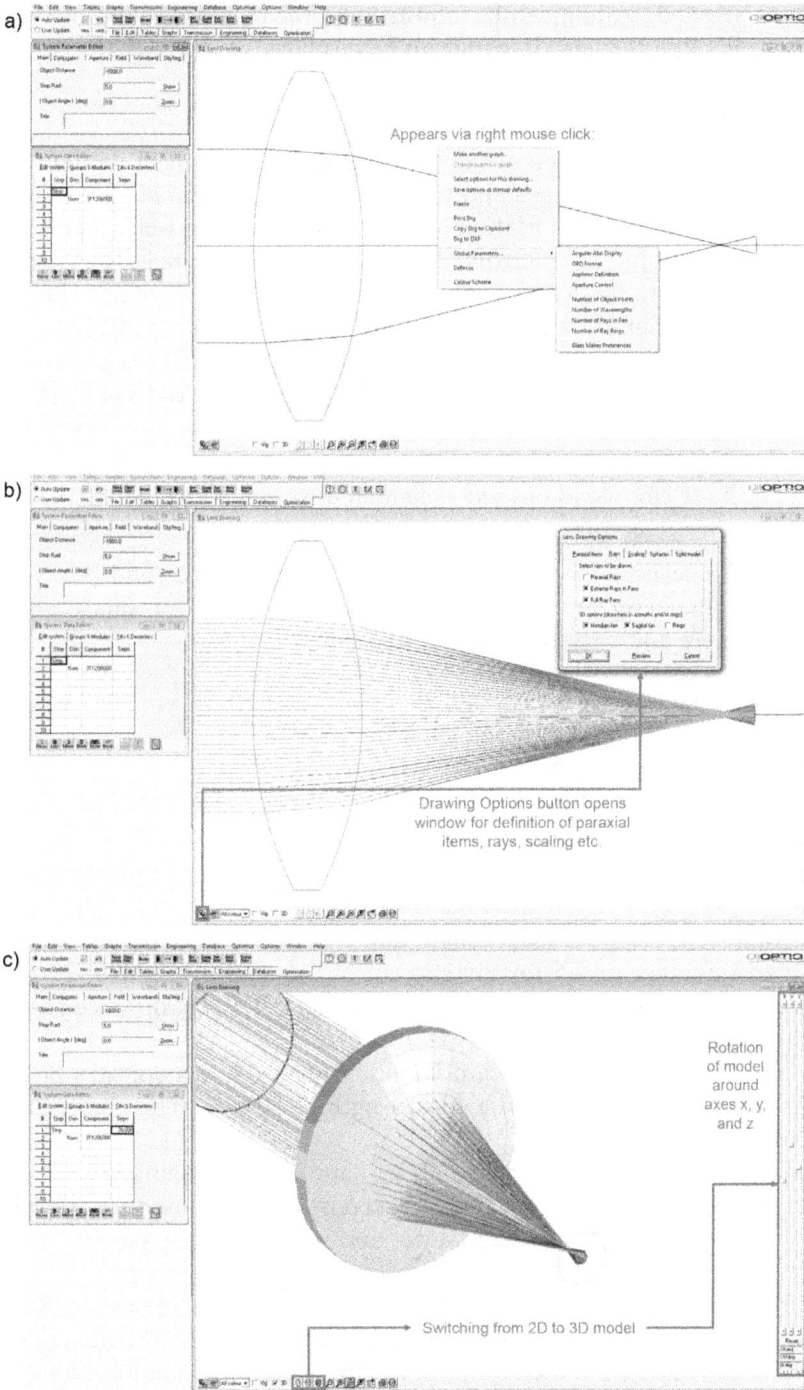

Figure 8.11. Presentation of different drawing options in WinLens, the relevant menus for adapting the number of rays/wavelength and other parameters (a, b) as well as the 3D view option (c).

rays can be selected, scaling can be adjusted and the displayed solid model can be customised. Finally, the lens drawing can be displayed in 2D and 3D.

8.2 Exercises

The following exercises represent typical tasks in optical system design. Different aspects, i.e. basic considerations and the choice and definition of start systems, the creation, evaluation and optimisation of lenses and systems as well as the simulation and impact of manufacturing errors and tolerances are covered. The exercises are arranged according to these topics. The detailed solutions to the particular exercises including additional information are provided in the appendix of this book. In some cases, there is not 'the one and only' solution for a given task. The solution presented in the appendix thus represents one practicable approach. And your approach or solution, dear reader, may be the better one sometimes...

8.2.1 Topic: basic considerations and definition of start systems

E1

Define any positive focal length, negative object distance and positive object height in PreDesigner. Now use the Slider function for quick adjustment provided here and vary the object distance...

E2

The following parameters are given:
- object distance = −1000 m
- object height = 200 mm
- image height = 6 mm
- aperture stop radius = 9 mm.

Enter the given parameters in PreDesigner. Which effect occurs...and why?

E3

Determine the required focal length and the type of appropriate start system with an f-number of 3.5 for imaging an object with a height of 20 mm on a detector where the image height is −2 mm and the object distance is −2000 m.

E4

An object with a height of 50 mm should be imaged by an optical system with a stop diameter of 20 mm and a focal length of 50 mm. The object distance is 120 mm.
   a. Determine an appropriate start system for this imaging task.
   b. Could the system determined in exercise 1 also be used for a 1:1 imaging of the object? If so, which object and image distance should be chosen?

E5

   a. An object with a height of 15 mm should be imaged on a super 8 film. The object distance is 2 m. The stop diameter of the used optical system is pre-defined and should amount to 40 mm. Determine the focal length of the required optical system.

b. Which optical system should be chosen?

c. Instead of air ($n_{air} = 1$), water ($n_{water} = 1.3$) is now the optical medium in the object space. Determine the change in focal length in this case.

E6

An object is imaged by an optical system with a stop diameter of 1' and a focal length of 150 mm where the object distance is 1 m. The object angle is 1°.

a. Which optical system can be used in this case?

b. Which system should be used when increasing the object angle to 20°?

c. What is the reason for the resulting differences?

E7

In an experimental setup, an object with a height of 4 mm has to be imaged on a detector. The object distance is 1 m and the stop diameter is set to 1'. There is only a single lens available. Its nominal focal length is 578.95 mm. You can choose between different cameras with different detectors:

1. CCD 1/10',
2. CCD 2/3',
3. CCD 1/2.5', or
4. TV 1/2'.

a. Which camera should be chosen?

b. Is the available single lens suitable for this imaging task?

E8

An object with a height of 5 mm is located at a distance of 1 m from an optical system with a stop diameter of 1'. The required magnification is −0.2

a. Determine the image height and distance.

b. Which optical system (focal length and type) shall be used in the present case?

c. Which optical system shall be used for an object with a height of 50 mm, 500 mm and 5 m, respectively when fixing the focal length and image distance, respectively?

E9

An object at infinity is imaged on a camera. The camera optics has a stop diameter of 20 mm and a focal length of 50 mm. The resulting image height is 16 mm. Compare the Airy disc diameter of this imaging case for different wavelengths, 200 nm, 550 nm and 3 μm.

E10

In order to evaluate the resolution of an optical system a calibration card (size: DIN A4-format) is imaged on a detector. The focal length of the optical system is 120 mm and its f-number amounts to 3.5. The magnification is −0.1. Determine the distance from the optical system where the calibration card should be placed in order to obtain good imaging quality.

E11

An object which is placed at infinity is imaged by an optical system with a stop radius of 5 mm and a focal length of 200 mm.

a. Determine the image height for this case.
b. Let us now apply the commonly-used approach of describing laser beams as perfectly collimated light rays. We use a red HeNe-laser and assume a laser beam radius of 5 mm (corresponding to the stop radius above). What is the resulting 'image height', i.e. the laser beam radius after focusing? Compare and discuss the differences of the values obtained in (a) and (b) on the basis of the fundamentals covered in chapter 2.

8.2.2 Topic: creating and evaluating lenses and systems

E12

Create the following lenses using pre-defined optical elements in the WinLens System Data Editor and determine the particular effective focal lengths:

a. Plano-convex, $R = 31.5$ mm, centre thickness $= 4$ mm, diameter $= 24$ mm, glass: E-LAK09 from Hikari.
b. Biconcave, $R_1 = 51$ mm, $R_2 = 20$ mm, centre thickness $= 2$ mm, diameter $= 30$ mm, glass: $n_d = 1.8052$.
c. Biconvex (best form), $R_1 = 125.75$ mm, $R_2 = 32.85$ mm, centre thickness $= 10$ mm, diameter $= 40$ mm, glass: N-BaK4 from Schott.

E13

a. Calculate the radius of curvature of a plano-convex lens with a diameter of 40 mm and a target focal length of 500 mm at a wavelength of 587 nm. The lens should be made of the glass 'N-BK7' from Schott and feature a centre thickness of 2 mm (here, a thin lens can be assumed). The index of refraction of the glass is 1.5168 at 587 nm.
b. Now create this lens in WinLens using a pre-defined element in the System Data Editor and compare the effective focal length given here with the default focal length mentioned in (a).
c. A collimated bundle of light rays is focused by this lens. The stop diameter is 38 mm and the aperture angle is $0°$ (i.e. the light rays arrive parallel to the optical axis). Which orientation of the lens should be. Should the plane surface be orientated towards the object or the image space?

E14

a. Create a biconvex lens with the following parameters: $R_1 = 50$ mm, $R_2 = 100$ mm, centre thickness $= 2$ mm, diameter $= 20$ mm and refractive index $n_d = 1.5$.
b. Determine the effective focal length of this lens and control the result provided by WinLens by calculating the focal length of a corresponding thin lens.
c. How does the focal length change if the lens is not made of the above-defined material, but of the glass N-SF1 from Schott...and why?

E15

Create a concave mirror with an effective focal length of -50 mm and a diameter of 40 mm.

a. We now consider an aperture diameter of 36 mm and a test wavelength of 546 nm. Determine the resulting Seidel sum S_I for spherical aberration.

b. Now reduce the aperture diameter to 18 mm and determine the corresponding Seidel sum S_I. Discuss the result with respect to the value observed for case (a).

c. Evaluate the chromatic aberration and discuss your observation.

E16

The following parameters are given: object distance: infinity, aperture diameter = 20 mm, wave band: visible (*d*, *C*, *F*, *s*, *g*).

a. Determine the longitudinal chromatic aberration (for 486.13 and 656.27 nm) of the catalogue lens with the order number 311 324.

b. Determine the longitudinal chromatic aberration (for 486.13 and 656.27 nm) of the catalogue achromatic doublet 322 308 and compare the result to the value found for case (a).

c. Discuss the different shapes of the curves displayed in the particular chromatic aberration diagram.

E17

Create a symmetric biconvex lens made of N-BK7 from Schott with a diameter of 20 mm. The radii of curvature amount to 100 mm and the centre thickness is 3 mm. Define the given parameters as follows: object at infinity, aperture stop radius = 8 mm, object angle = 10° and considered wave band = Fraunhofer lines *d*, *C*, *F*, *s* and *g*.

a. Increase the number of displayed (and thus actually calculated and traced) rays to 500 and increase the number of displayed object points from two (per default) to five.

b. Now take a look at the spot diagram. Which aberration can be seen here?

c. Increase the number of considered wavelengths in the spot diagram. Which additional aberration occurs now?

E18

An object shall be observed by a periscope where the illumination is white light. The object distance is 50 m. The periscope should feature a focal length of 1 m, a stop diameter of 80 mm and bridge a vertical height of 500 mm. Set up such a periscope consisting of two prisms and a single focusing optical component using standard catalogue optics.

E19

The following parameters are given:
- object height: 4.7 mm
- image height: 2 mm
- object distance: 500 mm
- stop diameter: 1′
- illumination: white light.

a. Determine the required focal length and type of optical system for this imaging task using the PreDesigner software.

b. Now transfer the given data to WinLens and insert an appropriate achromatic doublet from the catalogue database. The diameter of this doublet should be as close to the given stop diameter as possible. The wavelength of interest is 546 nm.

    c. Determine the resulting Seidel sums for spherical aberration and coma and open an appropriate graph for the visualisation of the resulting spot diameter as well as sagittal and tangential coma in the image plane.

E20

Create a convex best form lens with the radii of curvature $R_1 = 25.75$ mm and $R_2 = 34.21$ mm, a centre thickness of $t_c = 4.3$ mm and a diameter of 20 mm. The lens is made of the barium crown glass N-BaK1 from Schott.

    a. Determine the nominal effective focal length of this lens.

    b. We now consider an object distance of 200 mm and a centre wavelength of 1060 nm. Determine the resulting image distance.

    c. How does the image distance change if the lens is used as front lens for an underwater camera?

E21

The following parameters are given:

- object distance: infinity,
- aperture diameter: 12 mm, and
- wave band: visible (Fraunhofer lines d, C, F, s, g).

Create an optical setup consisting of an Amici prism (enter this as a pre-defined component and do not modify its dimensions, angles and glasses that are suggested by the Prism Editor) and a thin symmetric biconvex lens with a focal length of 100 mm and a diameter of 40 mm. The distance between the aperture stop and the prism should amount to 10 mm and the distance between the prism and the thin lens should be 20 mm. Now look at the spot diagram. What could this simple setup be used for?

E22

The following parameters are given:

- object distance: infinity,
- aperture diameter: 20 mm, and
- wave band: visible (d, C, F, s, g).

    a. Insert the achromatic doublet with the order number 322 352 000 in the System Data Editor and determine its longitudinal chromatic aberration (for 486.13 and 656.27 nm) using the chromatic aberration diagram.

    b. Now modify the diverging lens of the achromatic doublet by changing its material to the glass N-BK7 from Schott and determine the longitudinal chromatic aberration (for 486.13 and 656.27 nm) of the modified achromatic doublet.

    c. Discuss the observed differences.

E23

Create a cemented achromatic doublet with a diameter of 20 mm and the following parameters:

- Lens 1 (converging) $R_1 = 100$ mm, $R_2 = -80$ mm, $t_c = 3$ mm, glass: N-BK7 from Schott.
- Lens 2 (diverging) $R_1 = -80$ mm, $R_2 = -40$ mm, $t_c = 2$ mm, glass: N-SF66.

Now define the wave band 'visible d, C, F, s, g', a stop diameter of 18 mm, an object distance of 500 mm and an object angle of 5° and perform a sensitivity analysis for this imaging case. The aberrations of interest are spherical aberration and coma.

    a. Which surface of the achromatic doublet is the most critical one?

    b. Determine the corresponding Seidel coefficients for spherical aberration and coma of this surface (at $\lambda = 587.5618$ nm).

    c. Determine the radius of Petzval field curvature.

8.2.3 Topic: optimisation of optical systems

E24

An optical system with an image angle of 20° is applied for imaging an object on a CCD-chip. The pixels on this chip are quadratic and feature a lateral length of 14 μm. In order to achieve high resolution, each pixel should be illuminated separately by an image point. The ambient medium is air. The defect of interest is spherical aberration and the Seidel sum S_I of the actual optical system is 0.01. The upper and lower limit of this defect should be $d_{max} = 0.004$ and $d_{min} = 0.002$. Calculate the absolute value of the merit function.

E25

The optical system which was defined and evaluated in Exercise E19 (doublet, EFL ≈ 150 mm) should be optimised where the defect of interest is coma (no other defects are considered). The goal is to realise illumination of single pixels on a detector by one image spot where the pixel size is 1.7 μm. The ambient medium is air. The limits of coma should be $d_{max} = 0.000\ 04$ and $d_{min} = 0.000\ 01$. Calculate the absolute value of the merit function.

E26

In Exercise E5, a doublet with a focal length of 382.75 mm was identified as an appropriate start system for the imaging task defined there.

    a. Transfer the given parameters from Exercise E5 to WinLens in order to define the general conditions. The object and image space medium is air. Insert an appropriate lens into the System Data Editor using the catalogue optics database.

    b. Is the generated system diffraction-limited when defining a wavelength of interest of 546 nm?

E27

Create a biconvex lens made of F5-glass from Schott using a pre-defined element in WinLens. The lens parameters are $R_1 = 150$ mm, $R_2 = -101$ mm, centre thickness = 5 mm, and lens diameter = 1′. The following general conditions are given: aperture stop radius = 12 mm, object at infinity, wave band: VIS (Fraunhofer lines e, F', C', s and g).

    a. Determine the focal length of this lens.

    b. Determine the longitudinal chromatic aberration of this lens where the wavelengths of interest are 643.8469 and 479.9914 nm.

c. The goal is now to optimise the lens by reducing chromatic aberration without changing the lens geometry (radii, thickness). The focal length should not be altered significantly and the maximum acceptable change in focal length is 300 μm.

E28

Create a biconvex lens made of N-BK7-glass from Schott using a pre-defined element in WinLens. The lens parameters are $R_1 = 25$ mm, $R_2 = -50$ mm, centre thickness $= 10$ mm, and lens diameter $= 30$ mm. The following general conditions are given: aperture stop radius $= 14$ mm, object at infinity, wave band: VIS (Fraunhofer lines e, F', C', s and g).

    a. Determine the Seidel sum for spherical aberration at 546 nm.

    b. Now reduce the aperture stop radius from 14 to 13 mm and determine the resulting change in Seidel sum for spherical aberration.

    c. Now set the aperture stop radius back to 14 mm and define an object angle of 10°. Determine the resulting Seidel sum for coma.

    d. Now shift the stop position by 5 mm towards the object and determine the resulting change in Seidel sum for coma.

E29

The following general conditions are given: object distance $=$ infinity, aperture diameter $= 16$ mm, object angle $= 20°$, and wave band $=$ visible (Fraunhofer lines d, C, F, s, g). Create a cemented achromatic doublet with a diameter of 25.4 mm and the following parameters:

- Lens 1 (converging) $R_1 = 61.748$ mm, $R_2 = -44.348$ mm, $t_c = 6.25$ mm, glass: N-BK7 from Schott.
- Lens 2 (diverging): $R_1 = -44.348$ mm, $R_2 = -128.64$ mm, $t_c = 2.75$ mm, glass: N-SF5 from Schott.

    a. Determine the focal length of this achromatic doublet.

    b. Evaluate the distortion qualitatively via the appropriate field diagram. Which type of distortion occurs? Determine the Seidel sum for distortion.

    c. Now shift the stop by 2 mm towards the image space, what happens to the distortion (both qualitatively and quantitatively)?

    d. Determine the value of the merit function for the single defect distortion before and after shifting the stop position. The target value of the Seidel sum is $S_V = d_t = 0.001$ and the fault tolerance is $d_{tol} = 0.0005$.

    e. Determine the maximum deviation of image point coordinates from the desired position before and after shifting the stop position.

8.2.4 Topic: simulation of manufacturing errors and tolerances

E30

A plano-convex lens (made of N-BaK4 from Schott) has a centre thickness of 10 mm and a diameter of 36 mm. The curved surface has a nominal radius of curvature of 50 mm, but due to manufacturing errors, this surface is a toroid. It thus features two perpendicular planes with different radii of curvature. In the present case, the radius of curvature of the secondary plane is 25 mm.

a. Create such a lens using pre-defined optical elements in WinLens.
b. Now set the aperture diameter to 30 mm and place the object at infinity. Which optical aberration can be observed in the spot diagram?

E31

Insert a catalogue lens with the order number 311 318 000 into the System Data Editor and specify the following conditions in the System Parameter Editor:

- object distance: −200 mm
- stop radius: 10 mm
- distance from stop to lens: 50 mm
- wavelength band: visible (Fraunhofer lines e, C', F', s and g).

a. Determine the amount of peak-to-valley value (PV) of the transmitted wave front at a wavelength of 546 nm.
b. How does the PV value change if the lens is tilted by 5°?
c. How does the PV value further change if the lens is additionally decentred by 5 mm in the Y-direction?

E32

Create a lens with a diameter of 10 mm made of the glass LAC14 from HOYA where $R_1 = 30$ mm, $R_2 = -15$ mm and $t_c = 3$ mm. The wavelength of interest is 546 nm. Set the aperture stop radius to 4 mm, place the object at infinity and tilt the second lens surface by 10° around the X-axis. Determine the resulting change in PV and rms value of the transmitted wave front.

E33

Create a plano-convex lens with a radius of curvature of 50 mm, a thickness of 5 mm and a diameter of 1′. The lens should be made of a glass with an index of refraction of $n_e = 1.51$. Define the following parameters in the System Data Editor: the object should be placed at infinity, the stop diameter should be set to 22 mm, and the wavelength of interest is 546.074 nm.

a. Determine the impact of the lens orientation on imaging quality by means of the spot diagram and the wave front plot.
b. Determine the change in PV value of the wave front for both lens orientations if the first lens surface (either the plane or the curved one, depending on the orientation of the lens) is tilted by 0.1° around the X-axis.
c. Determine the change in PV value of the wave front for both lens orientations if the entire lens is tilted (without any tilt of the particular first lens surface) by 0.1°.
d. Determine the change in PV value of the wave front for both lens orientations if the entire lens is decentred by 100 μm in the Y-direction (i.e. a lateral offset from the optical axis).
e. Determine the change in PV value of the wave front for both lens orientations if the entire lens is decentred by 100 μm in Z-direction (i.e. a distance error with respect to the object and image space)?
f. Which error considered in this exercise has the highest impact on image quality—the orientation of the lens, the tilt of a single lens surface, the tilt of the entire lens, the lateral offset, or the distance error?

E34

Create a cemented lens doublet by defining two lenses with a common interface in the user-defined Lens Editor. The first lens should be a plano-convex lens with the following characteristics:

- $R_1 = 0$ mm
- $R_2 = -50$ mm (surface cemented with second lens)
- $t_c = 5$ mm
- diameter $= 1'$
- index of refraction $n_e = 1.51$.

The characteristics of the second lens, a meniscus, are:

- $R_1 = -50$ mm (surface cemented with first lens)
- $R_2 = -70$ mm
- $t_c = 3$ mm
- diameter $= 1'$
- index of refraction $n_e = 1.9031$.

a. Evaluate the impact of the orientation of this doublet using the spot diagram and the wave front plot.

b. Now tilt the first surface of the first lens by $0.1°$ in the X-direction and the second surface of the second lens by $0.1°$ in the Y-direction. Determine the impact of this tilt on the imaging quality using the spot diagram and the wave front plot.

c. Now redesign the doublet so that both lenses can be tilted and decentred independently from each other. How can this be realised?

d. Tilt the second lens by $0.5°$ around the Y-axis with respect to the first lens (which is not tilted at all). Evaluate the resulting impact on imaging quality using the spot diagram and the wave front plot.

e. The radii of curvature of both cemented surfaces have the following tolerance: 3/10(2/0.1) where the test wavelength is 633 nm. The used cement has an index of refraction of 1.6514; the centre thickness of the cement layer is 200 μm. Determine the difference in the wave front's PV value for the maximum form deviation or error that can arise at the cemented surfaces.

References

[1] Gerhard C and Adams G 2015 *Proc. SPIE* **9793** 97930N
[2] Thöniß T, Adams G and Gerhard C 2009 *Opt. Photonik* **4** 30–3
[3] Adams G, Thöniß T and Gerhard C 2013 *Opt. Photonik* **8** 50–3

IOP Publishing

Lens Design Basics
Optical design problem-solving in theory and practice
Christoph Gerhard

Appendix A

A.1 Solutions of exercises

A.1.1 Topic: basic considerations and definition of start systems

E1

We see that when changing the object distance, the image distance is adapted in real time since it is directly conjugated to the object distance as expressed by the imaging equation. Moreover, the object and image height change since both heights depend on the distances via the magnification.

E2

After entering the given parameters in PreDesigner, a virtual image is formed. Such a virtual image occurs since both the object height and the image height were entered as a positive value. By changing the algebraic sign of one height, we obtain a real image.

E3

For the given parameters, a promising start system is a doublet with a focal length of 181.82 mm according to the calculation via PreDesigner.

E4

    a. After entering the given parameters in the corresponding boxes in PreDesigner and displaying the extra information, a double Gauss lens is suggested as an appropriate start system for the given imaging task as shown in figure A1.

    b. In order to obtain 1:1 imaging (i.e. object height = image height), the image height instead of the object distance has to be defined. It corresponds to the object height of 50 mm. The object distance thus corresponds to the image distance where the absolute value is 100 mm as shown in figure A2. Here, the same system as in part (a) of this exercise, a double Gauss lens, is suggested as the start system.

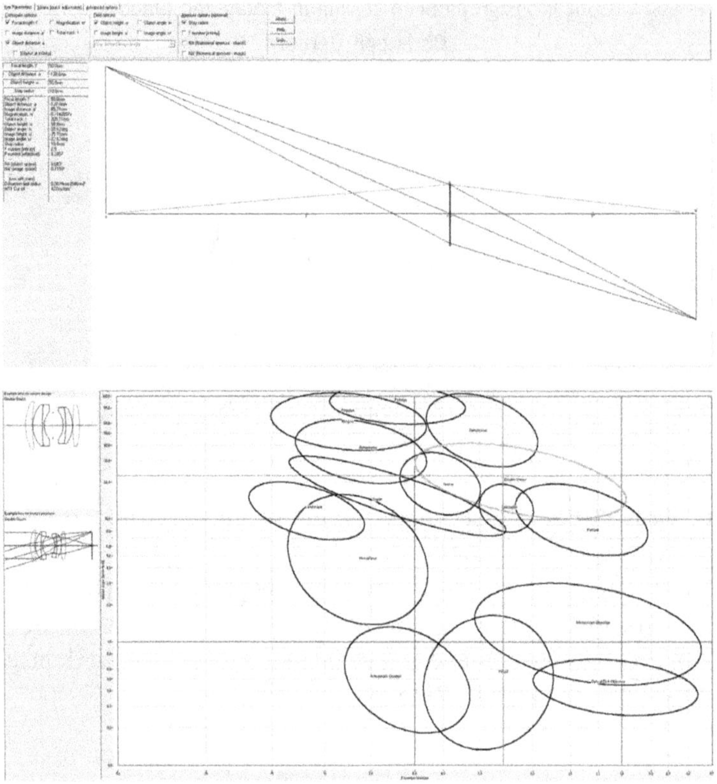

Figure A1. For the given parameters (top) a double Gauss lens is suggested as a start system (bottom).

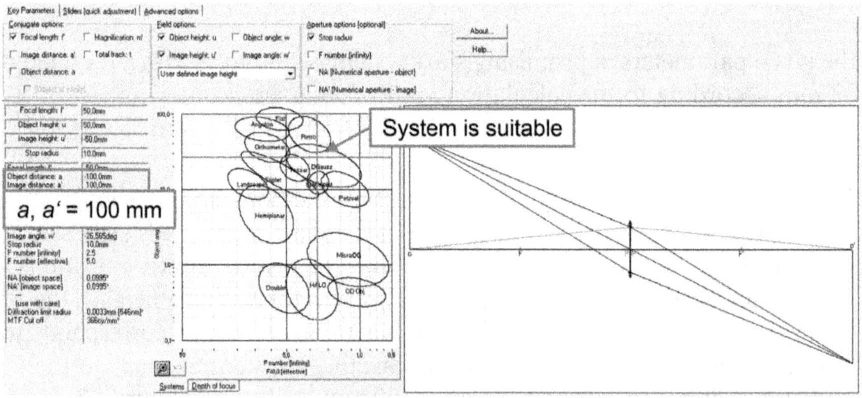

Figure A2. The system is suitable for 1:1 imaging.

a)

Back focal length BFL = f' = 382,75 mm

b)

c)

$f' = 439,53$ mm $\rightarrow \Delta f \approx 57$ mm

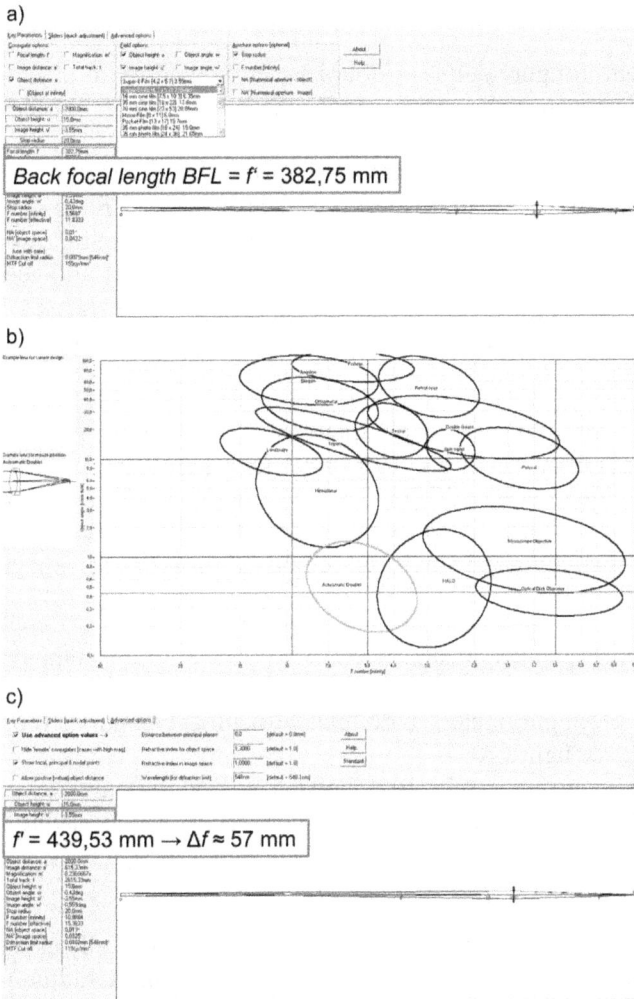

Figure A3. Determination of the focal length, the start system and the change in focal length due to a change in the index of refraction in the object space.

E5

    a. After entering the given parameters in the corresponding boxes in PreDesigner, a focal length of 382.75 mm is calculated, see figure A3(a).

    b. The suggested start system is a doublet as shown in figure A3(b).

    c. The index of refraction in the object space can be changed in the Advanced Options function[1], see figure A3(c).

---

[1] In this function, the indices of refraction in the object and image space, the wavelength of interest and a number of other parameters and settings can be defined.

E6

    a. As shown in figure A4(a), a doublet can be used at an object angle of 1°.

    b. When increasing the object angle to 20°, a triplet should be used, see figure A4(b).

    c. Obviously, the different start systems result from the difference in object angle. At an object angle of 1° the field of view is marginal and since the angle is lower than 5°, the approach of paraxial imaging can be applied. In contrast, an object angle of 20° exceeds the limit of 5° and geometric-optical imaging has to be applied where the object angle has to be taken into account. In this context, a triplet allows for the reduction of distortion etc.

E7

    a. When entering the given parameters in PreDesigner it turns out that the resulting image height is −5.5 mm, see figure A5(a). This image height corresponds to a 2/3′ CCD-chip; the second camera should thus be chosen.

    b. As shown in figure A5(b), both the object angle and the f-number are quite low. In such a case, a single lens could be sufficient for imaging.

E8

    a. The image height and distance are −1 mm and 200 mm, respectively, see figure A6.

    b. For the given parameters, a doublet with a focal length of 166.67 mm should be used, see figure A6.

    c. As shown in figure A7, the following optical systems are suggested when increasing the object height u (or y) successively: a hemiplanar for $u = 50$ mm, a triplet or orthometar for $u = 500$ mm and a fisheye for $u = 5$ m.

E9

When entering the given parameters the software automatically calculates the Airy disc radius (and not the diameter!), a.k.a. diffraction limit radius for the default

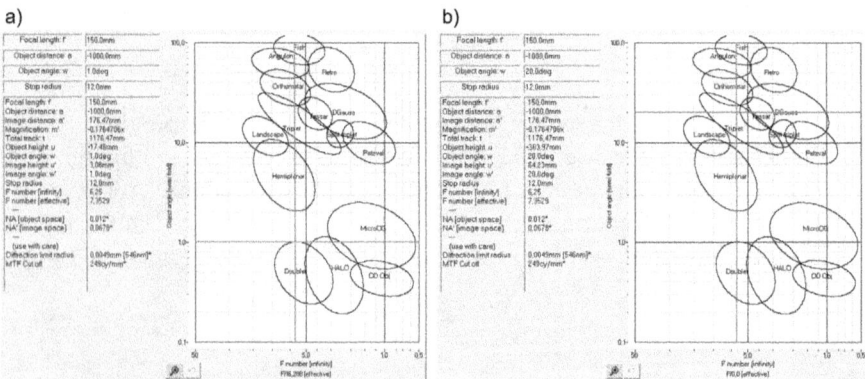

Figure A4. Choice of an appropriate start system at an object angle of 1° (a) and 20° (b).

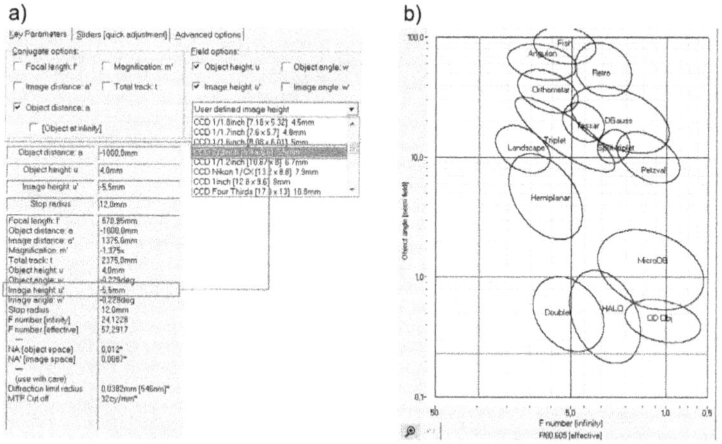

Figure A5. Choice of the camera on the basis of the resulting image height (a) and selection of the start system (b)—where no system is suggested due to the low object angle and f-number.

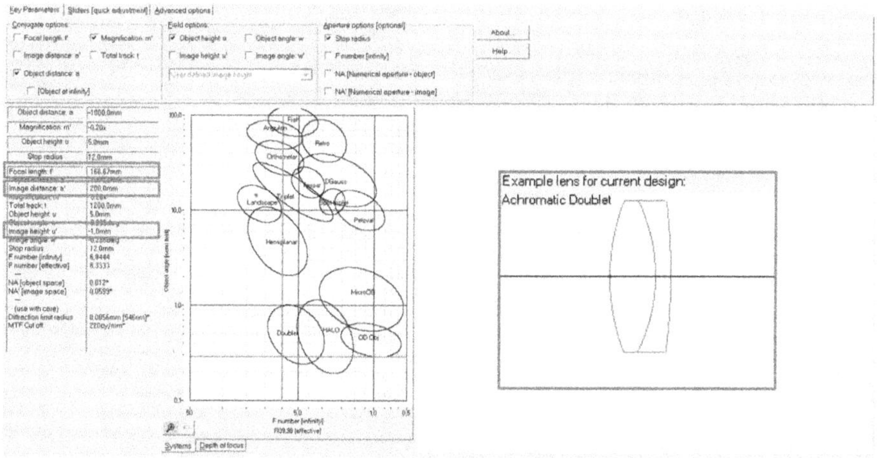

Figure A6. Determination of the image height and distance as well as the start system.

wavelength 546 nm as shown in figure A8(a). The wavelength can be changed in the Advanced Options function, see figure A8(b). For the given wavelengths, the Airy disc diameters are 1.2 μm at 200 nm, 3.4 μm at 550 nm and 18.4 μm at 3 μm.

E10

Here, merely the format of the calibration card is given. Its corresponding object height thus has to be determined by the following basic consideration: the size of a DIN A4-object is 210 mm × 297 mm. The object diagonal c is then

$$c = \sqrt{(210 \text{ mm})^2 + (297 \text{ mm})^2} = 363.74 \text{ mm}.$$

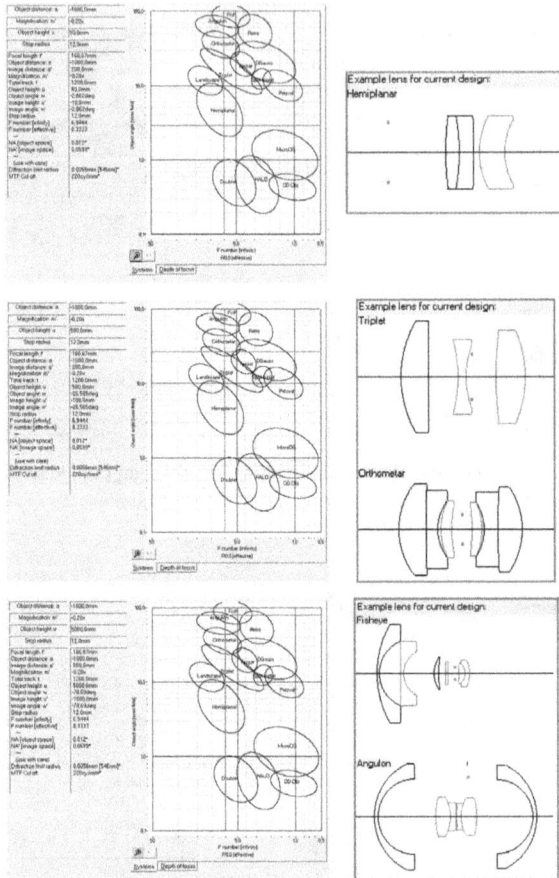

Figure A7. Suggested start systems for different object heights, $y = 50$ mm (top), 500 mm (middle) and 5000 mm (bottom).

The object height is thus half the diagonal, $c/2 = 181.87$ mm ≈ 182 mm. After inserting the given parameters and the calculated object height, an object distance of 1320 mm is determined by PreDesigner.

E11

a. As expected and shown in figure A9(a), the image height u' (or y') is 0 mm since in the ray optical model, it is given by the infinite small intersection of two lines[2].

b. In PreDesigner, the mode of calculation can be switched from ray optics to Gaussian beam propagation by selecting the Gaussian beam function as shown in figure A9(b). After setting the given laser wavelength (red HeNe-laser

---

[2] ...even though the wave-optical diffraction limit, i.e. the Airy disc diameter, can be determined.

a)

b)

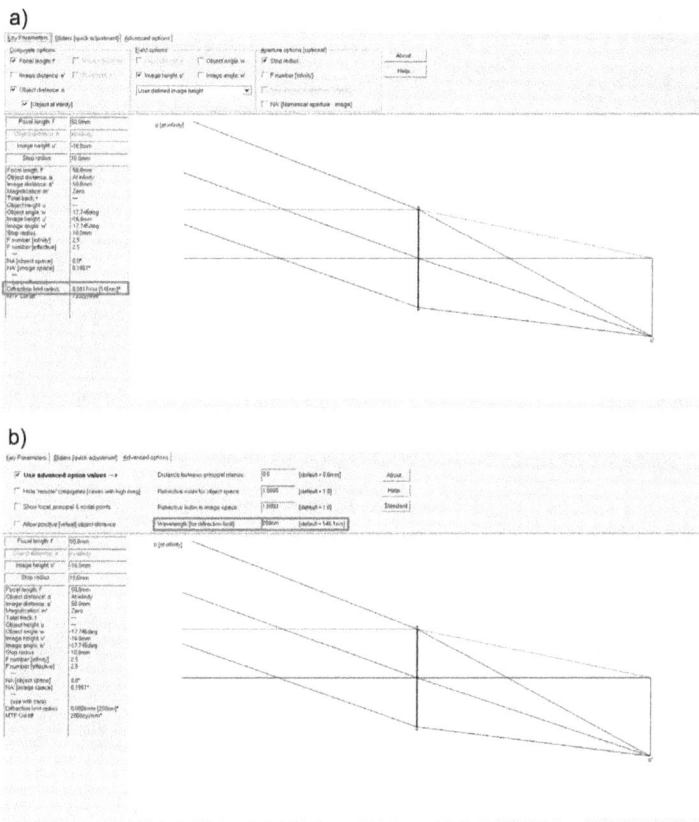

Figure A8. Determination of the Airy disc radius (a) and changing the wavelength of interest (b).

wavelength = 632.8 nm) a beam radius in the image space of $r' = 8.1$ μm is calculated. The laser beam waist diameter of the rela beam after focusing is thus 16.2 μm.

A.1.2 Topic: creating and evaluating lenses and systems

E12

The particular effective focal lengths are:
 a. 42.72 mm,
 b. −17.46 mm, and
 c. 46.66 mm.

These values are calculated based on the lensmaker's equations and displayed during the setup of a lens as shown in figure A10. Note that for choosing the given glasses, the boxes 'Available?' and 'Obsolete' should be checked in the glass finder since per default, merely recommended glasses are displayed here.

a)

$u' = y' = 0\ \mu m$

b)

Set laser wavelength to 632.8 nm (HeNe)

$r' = 8.1\ \mu m \neq 0$

Use *Gaussian Beam* function

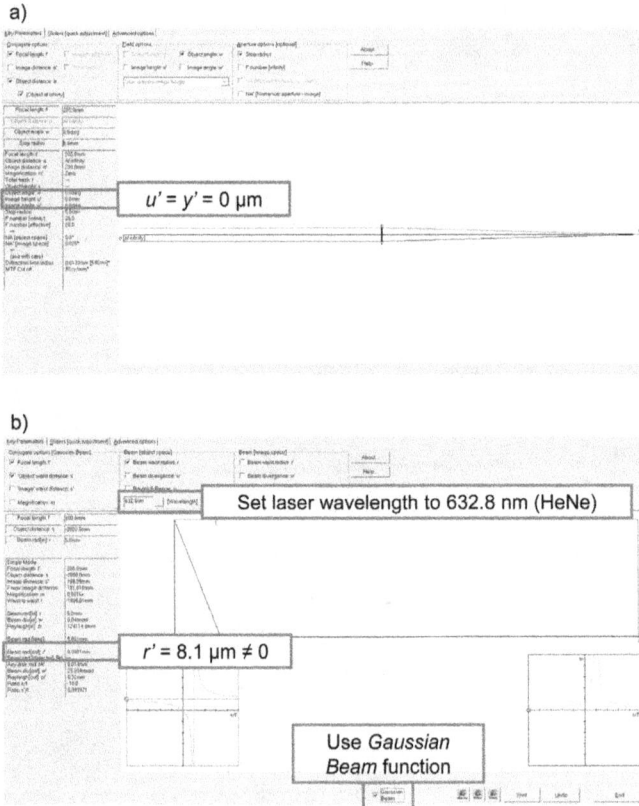

Figure A9. Using the Gaussian beam function for comparing paraxial calculations (a) to wave-optical calculations (b).

E13

   a. The radius of curvature can be derived from the lensmaker's equation of a thin plano-convex lens given by equation (3.3). It amounts to

$$R = EFL \cdot (n - 1) = 500\ mm \cdot (1.5168 - 1) = 258.4\ mm.$$

   b. The lens created by a pre-defined element in the system data editor features a focal length of 500 mm and thus confirms the result calculated in (a). Actually, this is not surprising since both approaches, the manual and the computer-assisted methods, are based on the same mathematical interrelationship, the lensmaker's equation.

   c. After the definition of the object distance (= infinity) and the stop radius (= 19 mm) in the System Parameter Editor the impact of the lens orientation on imaging quality can be rated. For instance, this can be carried out via the spot diagram. As shown in figure A11, a better result in terms of size and position of the image is obtained if the curved lens surface is orientated towards the object. In this case, the angles of incidence and refraction,

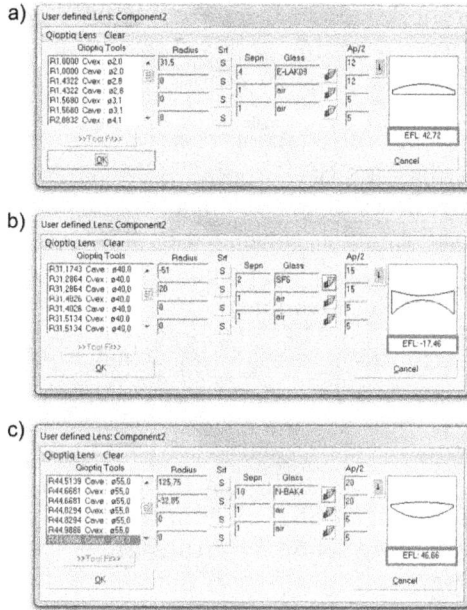

Figure A10. Lens editor showing the preview of the lens design and its effective focal length.

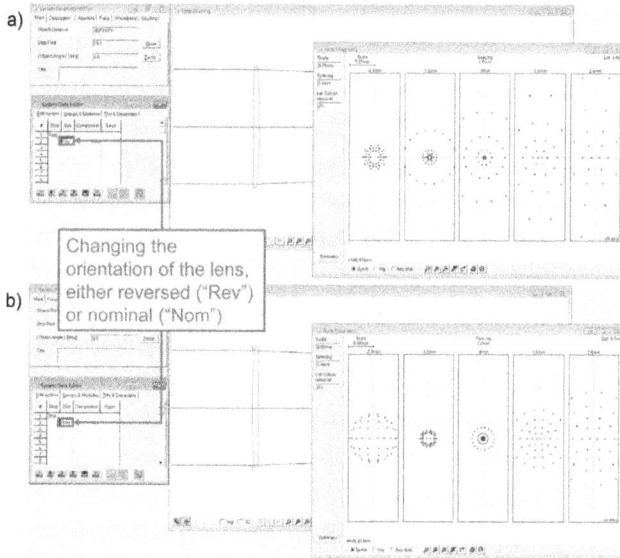

Figure A11. Comparison of the impact of lens orientation on the spot diagram with (a) plane surface towards object and (b) curved surface towards object.

respectively, on both lens surfaces are minimised, resulting in a minimisation of spherical aberration.

E14

a. A glass with an index of refraction of $n_d = 1.5$ is not available in the glass database. Thus, the glass HK2 from CDGM can be chosen since it features the least deviation from the target value.

b. The effective focal length as calculated by WinLens after inserting the radii of curvature and lens thickness amounts to 66.66 mm. The same value is obtained when applying equation (3.1):

$$\text{EFL} = \frac{1}{n-1} \cdot \left(\frac{R_1 \cdot R_2}{R_2 - R_1} \right) = \frac{1}{1.5 - 1} \cdot \left(\frac{20 \text{ mm} \cdot (-100 \text{ mm})}{(-100 \text{ mm}) - 50 \text{ mm}} \right) = 66.66 \text{ mm}.$$

c. When changing the glass, the effective focal length decreases from 66.66 mm (for HK2) to 42.34 mm (N-SF11). The difference is thus 24.32 mm due to the fact that the index of refraction is increased from 1.5005 (for HK2) to 1.7847 (for N-SF11).

E15

a. The effective focal length of a mirror is half its radius of curvature, i.e. −100 mm in the present case. For the given parameters, the Seidel sum for spherical aberration amounts to $S_I = 0.209\ 95$.

b. After reducing the aperture diameter from 36 to 18 mm the first Seidel sum accounts for $S_I = 0.013\ 12$. This corresponds to a decrease by a factor of 16 due to the shadowing of the outer section of the incoming bundle of light rays. Consequently, the ray entrance height and angle of incidence or refraction of light rays, respectively, is decreased, finally reducing the effect of spherical aberration.

c. Chromatic aberration can be evaluated via the chromatic aberration diagram. Moreover, WinLens provides a kind of Seidel sum for both longitudinal and lateral chromatic aberration, CI and CII, respectively[3]. These values can be accessed via the Seidel aberrations table. As shown in figure A12, no chromatic aberration occurs. There is no chromatic aberration since a mirror is a reflective, but not a diffractive optical component. Hence, light does not pass through any optical medium and is thus not subject to dispersion.

E16

a. The longitudinal chromatic aberration can be determined via the chromatic aberration diagram where the effective focal length for the chosen wavelengths that define the wave band are displayed. For the two wavelengths of

---

[3] CI and CII are not true Seidel sums since there are only five officially existing Seidel aberrations. However, these values are quite helpful for the evaluation of optical systems.

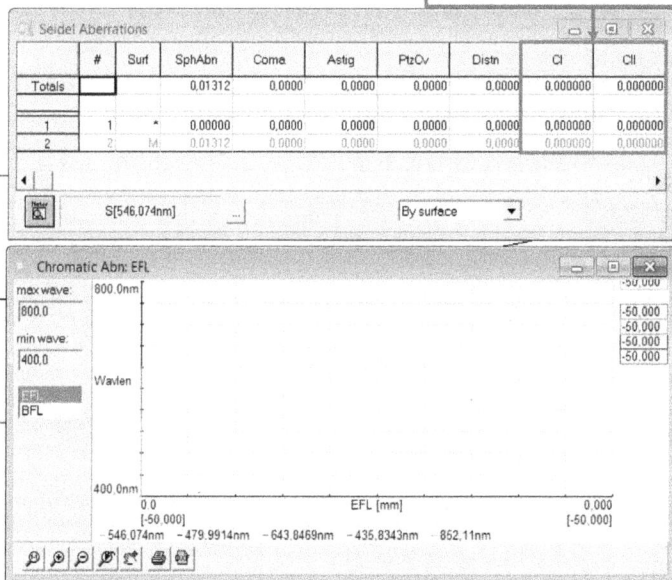

Figure A12. For mirrors, no chromatic aberration is detected.

interest, the catalogue lens with the order number 311 324 000 features a difference in effective focal length—and thus a longitudinal chromatic aberration—of $101.223 - 99.674 = 1.549$ mm, see figure A13.

b. For the achromatic doublet with the order number 322 308 000, this value amounts to $100.260 - 100.178 = 0.082$ mm $= 82$ μm. It is thus approximately 19-times lower than for the single lens even though both considered elements feature the same effective focal length.

c. The curve shown in figure A13(a) directly follows from the dispersion characteristic of the lens material: the higher the wavelength, the longer the effective focal length[4]. The curve shown in figure A13(b) represents the so-called secondary spectrum, which occurs after the correction of chromatic aberration for two wavelengths.

E17

a. The number of displayed rays can be changed in the Global Parameters function that can be accessed via a right mouse click on any evaluation graph or the lens drawing as shown in figure A14. The required number of 500 rays

---

[4] Principally, the function EFL(λ) is the mirror inverted function $n(\lambda)$, i.e. the dispersion curve.

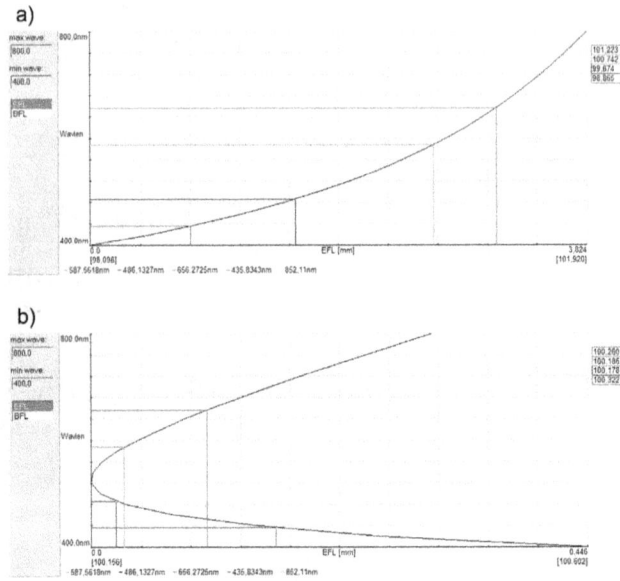

Figure A13. Comparison of the chromatic aberration diagram of (a) a single lens and (b) an achromatic doublet; the nominal effective focal length of both elements is 100 mm.

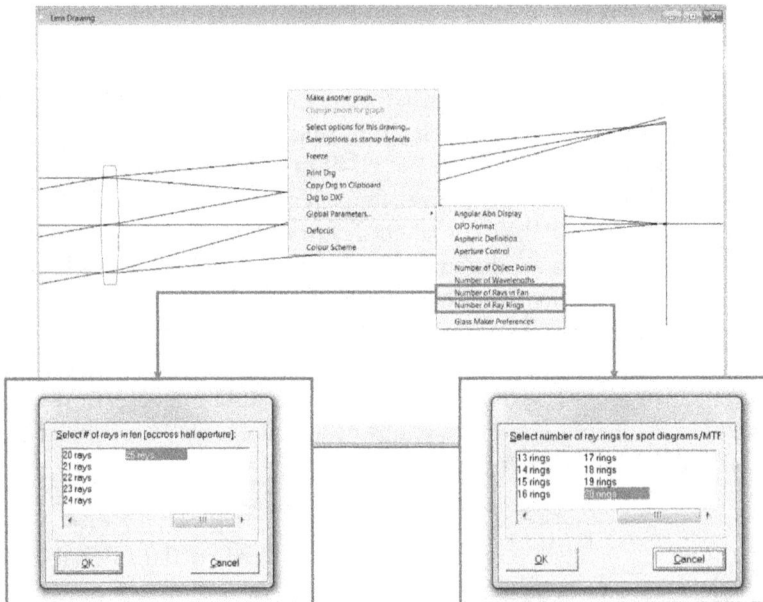

Figure A14. Changing the number of traced and displayed rays using the Global Parameters function.

is achieved by choosing, for example, 25 rays in a fan and 20 ray rings (25 × 20 = 500). The number of object points can be changed in the same function.

b. The spot diagram clearly indicates coma.

c. After increasing the number of displayed wavelengths via the Global Parameters function, chromatic aberration is displayed additionally in the spot diagram—as expected.

E18

The periscope should consist of a focusing element and two prisms. As a start, the focusing element can be selected from the database. Its focal length can be chosen arbitrarily, but should at least be higher than the vertical height. In order to avoid image blurring due to spherical aberration, a doublet instead of a single lens can be chosen. One approach (but not the only one) is thus to select an achromatic doublet with a focal length of 1 m, for example, the catalogue doublet with the order number 322 241 000. The diameter of this doublet is 80 mm, i.e. the required stop diameter. Once this doublet is added to the System Data Editor, a 90° reflective prism can be selected from the database (e.g. order number 339 921). When adding such a prism, the optical axis is automatically folded and folded again after adding the second prism as shown in figure A15. Finally, the dimensions of the prism and the distances between the optical elements can be adjusted if necessary, see figure A16.

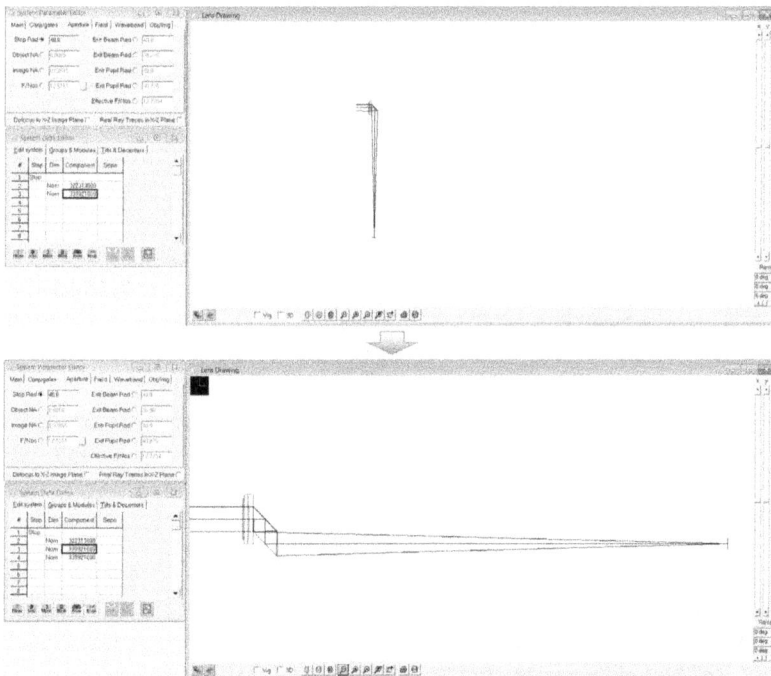

Figure A15. Inserting the components needed for setting up a simple periscope.

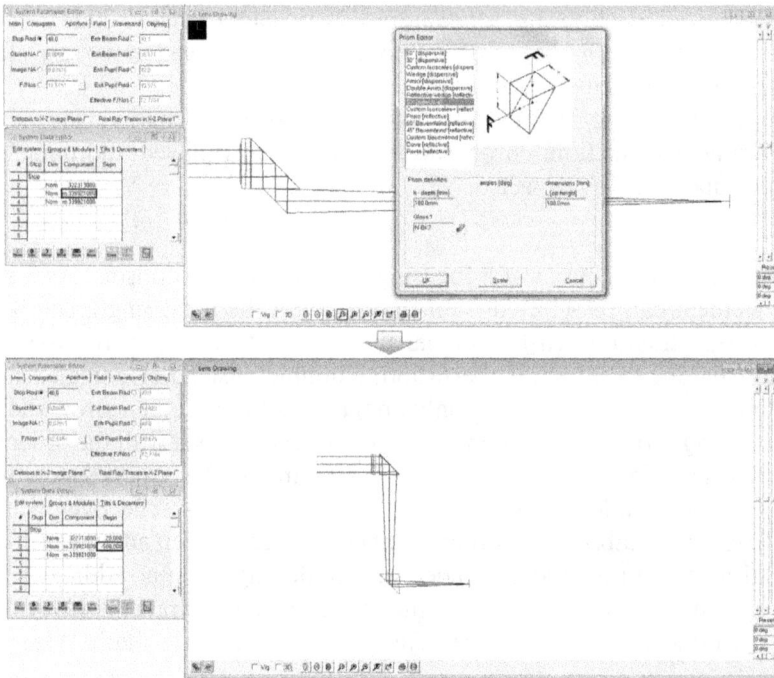

Figure A16. Changing the dimension of a prism and adjusting air gaps by shifting optical components.

E19

   a. After entering the given parameters in the corresponding boxes in PreDesigner a focal length of 149.25 mm is determined. An appropriate start system is identified via the extra information, see figure A17. It is a doublet.

   b. In WinLens, the object distance, the stop radius, the object size and the wave band are defined in the System Parameter Editor, see figure A18. The doublet is chosen from the catalogue database (here, an achromat = achromatic doublet is selected) as shown in figure A19. Its effective focal length is 150 mm and the diameter amounts to 31.5 mm ('as close to the given stop diameter as possible').

   c. The value of the particular Seidel sum can be accessed via the Seidel Aberrations table as shown in figure A20(a). For spherical aberration, it is $S_I = 0.0058$ and for coma, it amounts to $S_{II} = -0.0006$. The spot diameter as well as the sagittal and tangential coma in the image plane can be visualised and evaluated by the spot diagram, see figure A20(b).

E20

   a. The nominal effective focal length of the lens is 26.35 mm.

   b. The centre wavelength of the pre-defined wave band 'IR Schott' is the required wavelength $\lambda = 1060$ nm, see figure A21(a). After selecting this wave band and entering an object distance of −200 mm, the resulting image distance is 29.42 mm as shown in figure A21(b).

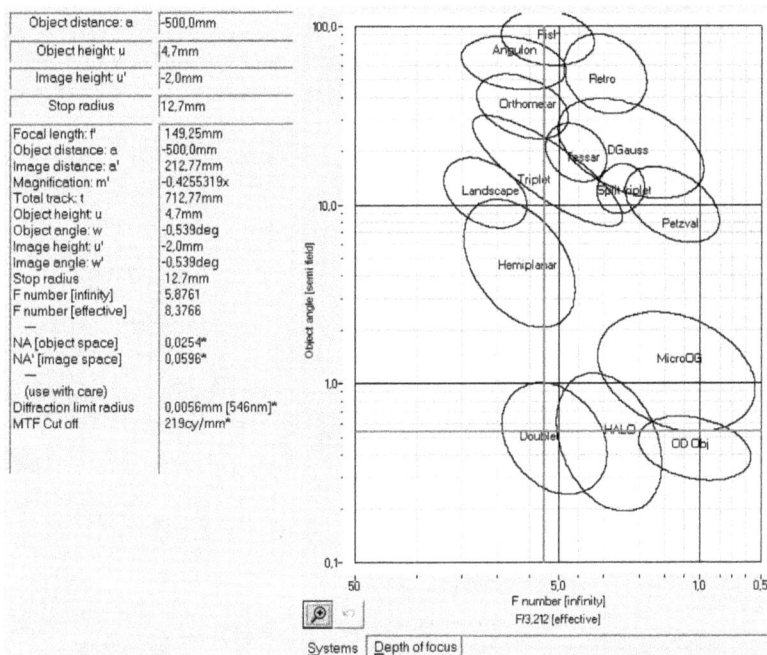

Figure A17. Determination of the adequate start system—an achromatic doublet.

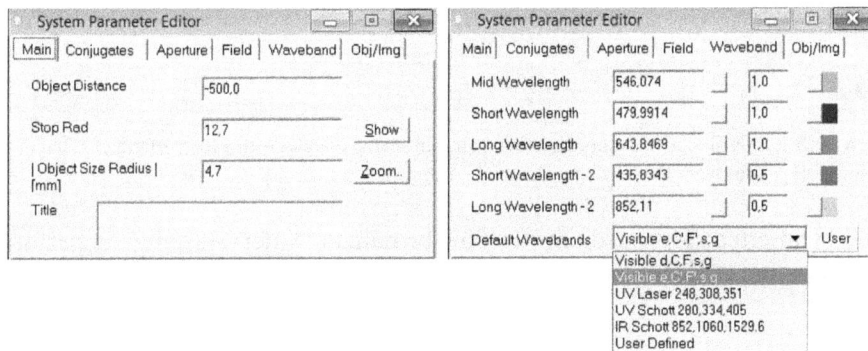

Figure A18. Definition of the object distance, the stop radius, the object size and the wave band in the System Parameter Editor.

c. When changing the index of reaction of the surrounding medium in the object space as shown in figure A21(c), the 'new' image distance[5] amounts to 31.07 mm, see figure A21(d). Here, an index of refraction $n_{water} = 1.33$ has to

---

[5] This effect, a change in image distance and the related or conjugated parameters, can be observed during diving or swimming. The human eye is 'designed' for the 'standard' surrounding medium, air. Under water, the basic conditions for this design are not valid, leading to an improvement or a degradation of the imaging quality—depending on the particular eye. In some cases, even defective vision disappears under water. For other cases, goggles with adequate optical power are available.

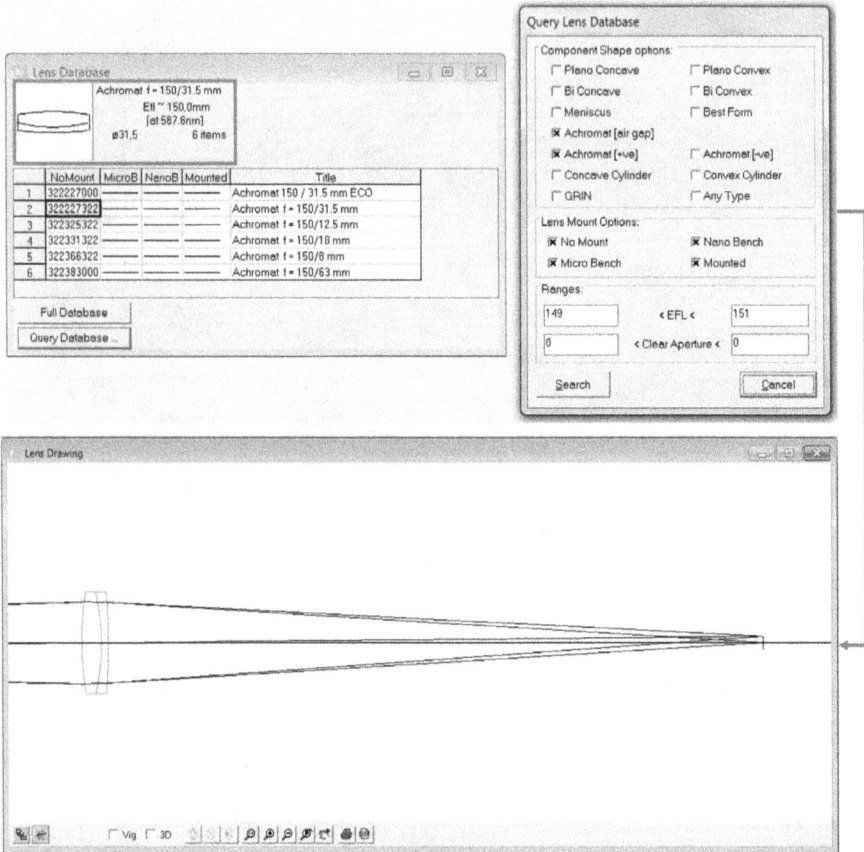

Figure A19. Selecting an optical component (here: an achromatic doublet with a focal length of 150 mm and a diameter of 31.5 mm) from the catalogue database.

be inserted. This medium can now be named 'water'—as already performed in figure A21(c).

E21

Inserting an Amici prism causes problems as shown in figure A22. According to the error message, the scale is too large to draw the entire lens. This is due to the fact that no focusing element was added to the design. The computer thus tries to calculate imaging from minus infinity (object distance) to plus infinity (image distance). Obviously, such an impossible calculation leads to a system crash. The solution is thus to insert the thin lens prior to the prism. For this purpose, the pre-defined element can be chosen where merely the effective focal length, the clear radius (i.e. half the lens diameter) and the lens bending (here: 0 = equi-curved = symmetric biconvex) have to be defined. The result after adjusting the distances or air gaps including the spot diagram is shown in figure A23. We see that the five wavelengths are clearly separated on the image plane and the detector, respectively. The setup could thus be used as a simple spectrometer.

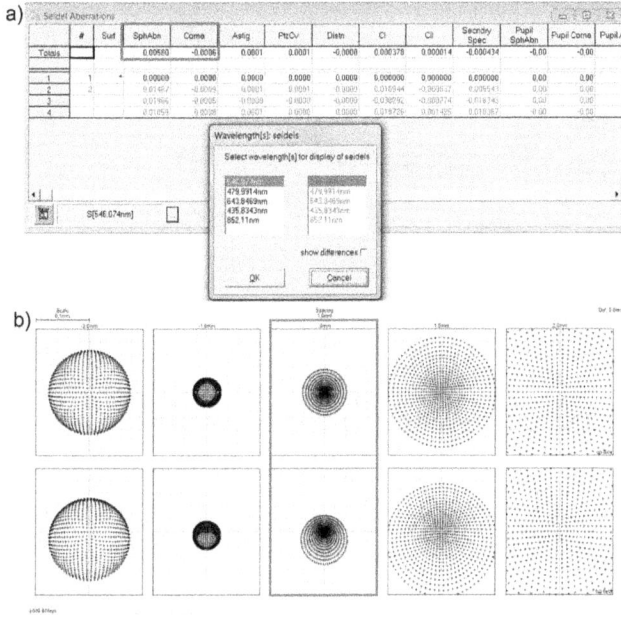

Figure A20. Evaluation of the formation and intensity of spherical aberration and coma via the Seidel Aberrations table (a) and the spot diagram (b).

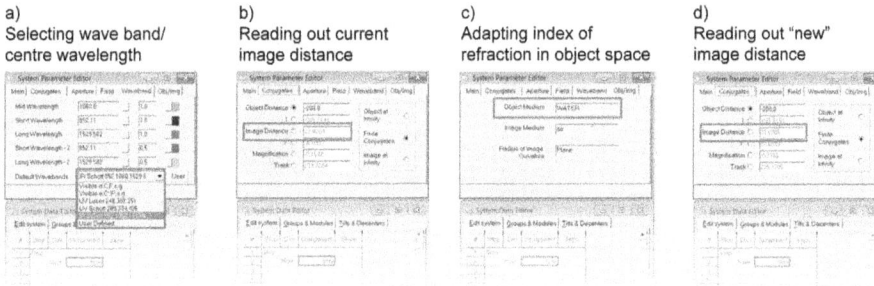

Figure A21. Selecting the considered wave band (a), adapting indices of refraction (c) and reading out wanted parameters (b and d) in the System Parameter Editor.

E22

a. According to the chromatic aberration diagram the longitudinal aberration of the given achromatic doublet is 160.126 − 160.034 = 0.092 mm = 92 μm.

b. After changing the glass of the diverging lens, the longitudinal chromatic aberration amounts to 131.253 − 129.242 = 2.011 mm, i.e. approximately 22-times higher than for the original achromatic doublet.

c. This difference is due to the fact that after changing the glass of the diverging lens, the condition for achromatism is no longer fulfilled as summarised in

Inserting single prism → error message!

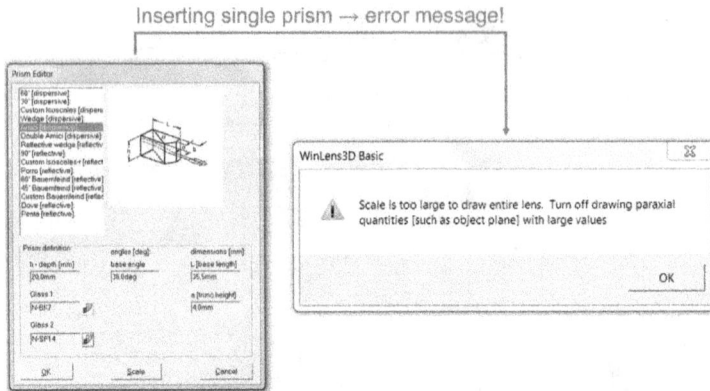

Figure A22. Inserting a single prism leads to an error message.

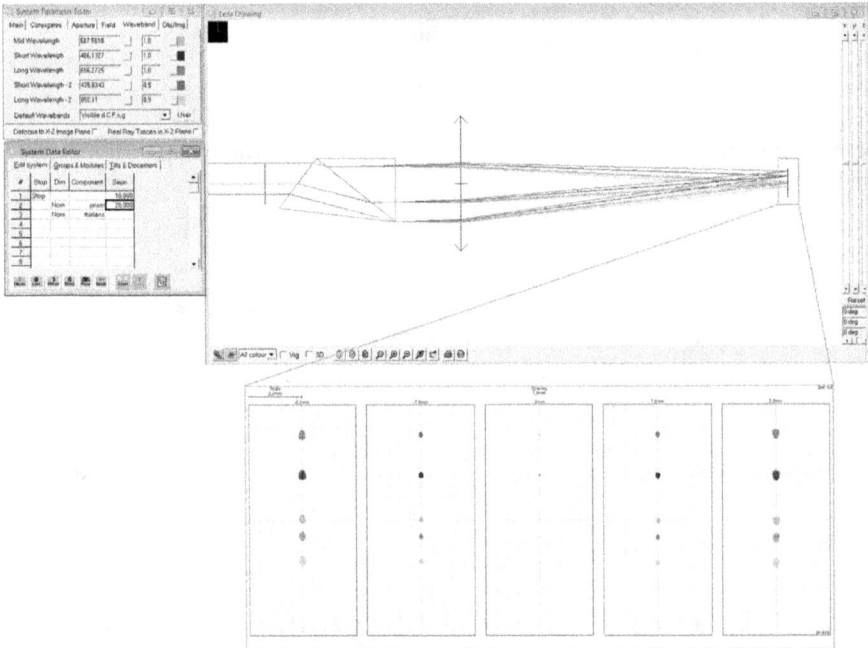

Figure A23. A simple spectrometer consisting of an Amici prism and a single thin lens. The spot diagram shows the distinct separation of image spots of different colour or wavelength.

figure A24. In other words: since both lenses are now made of the same glass, the achromatic doublet was converted into a thick converging lens.

E23

The created achromatic doublet (EFL = 43.52 mm) and the corresponding Seidel bar chart for sensitivity analysis are shown in figure A25(a). Here, the particular bars for the aberrations of interest, spherical aberration and coma, are displayed.

Before modification:

N-BK7 (V = 64.17) + N-SF5 (V = 32.25)

$EFL_1 \cdot V_1 =$
80.89 mm · 64.17 =
5,190.71 mm

$EFL_2 \cdot V_2 =$
-161.92 mm · 32.25 =
-5,221.92 mm

After modification:

N-BK7 (V = 64.17) + N-BK7 (V = 64.17)

$EFL_1 \cdot V_1 =$
80.89 mm · 64.17 =
5,190,71 mm

$EFL_1 \cdot V_1 =$
-210.51 mm · 64.17 =
-13,508.43 mm

Condition for achromatism: $EFL_1 \cdot V_1 = -EFL_2 \cdot V_2$

5,190.71 mm ≈ 5,221.92 mm   5,190.71 mm ≠ 13,508.43 mm

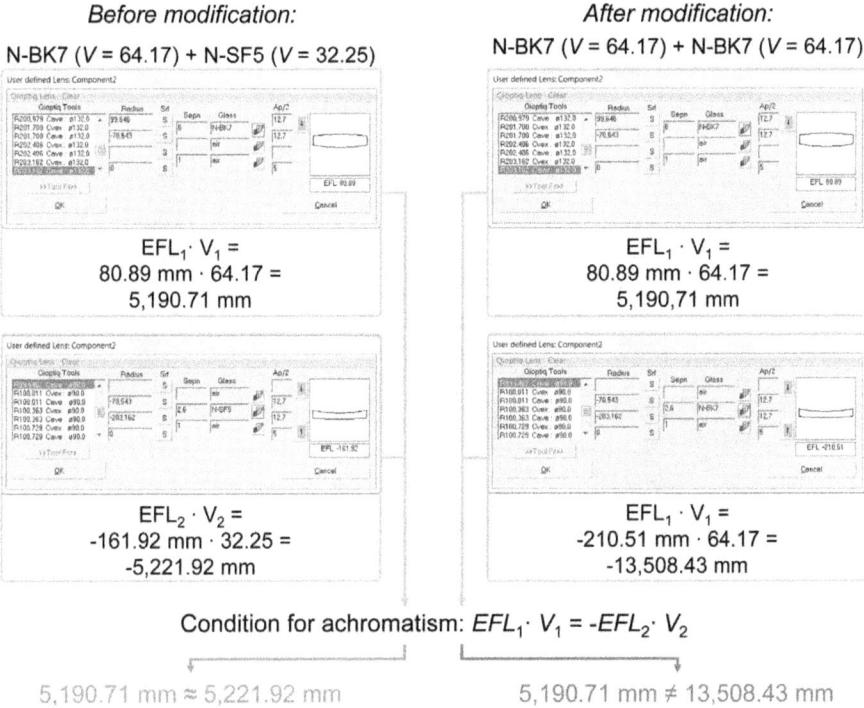

Figure A24. Comparison of the achromatic doublet before and after modification of the material of the second lens by means of the condition for achromatism.

a. It can clearly be seen that the last surface, i.e. the rear surface of the doublet, is the most critical one. Since surface number one is the stop position, it is surface number four.
b. The Seidel coefficients for spherical aberration and coma at this surface are 0.298 47 and 0.0545, respectively, see figure A25(b).
c. The radius of Petzval field curvature can be displayed in the field diagram for astigmatism as shown in figure A25(c). In the present case it amounts to −73.177 mm.

A.1.3 Topic: optimisation of optical systems

E24
The following parameters are given: image angle $u' = 20° = 0.35$ rad[6], $n = 1$ (since the medium is air), $l_{pixel} = 14$ µm $= 0.014$ mm[7], $d_{max} = 0.004$ and $d_{min} = 0.002$, and $d_a = 0.01$. The merit function, given by

---

[6] For the calculation of Seidel sums, the angle has to be given in radian. The corresponding unit 'rad' is usually not denoted.

[7] The lateral pixel size corresponds to the maximum spot diameter if one image point should be imaged on one pixel as requested. For the calculation of Seidel sums, this value is given in millimetres, but treated as a dimensionless quantity without any unit.

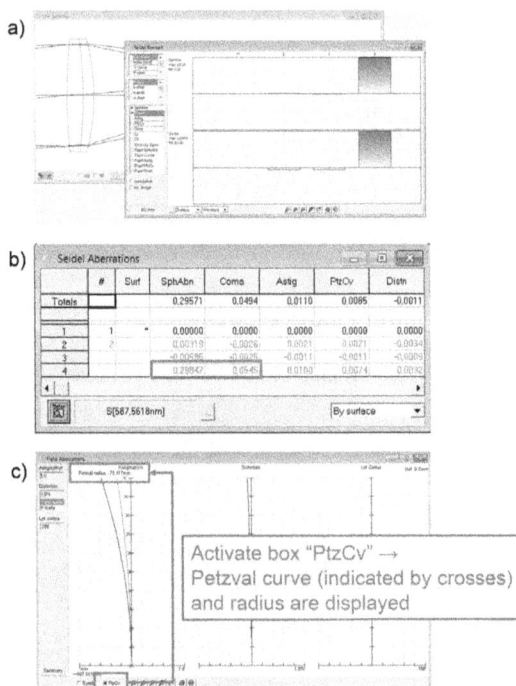

Figure A25. Seidel bar chart (a), Seidel Aberrations table (b) and field diagram for astigmatism (c) of the created doublet.

$$\text{MF} = \sum_i d_{i,\text{rel}}^2 = \sum_i \left(\frac{d_{i,\text{a}} - d_{i,\text{t}}}{d_{i,\text{tol}}} \right)^2 = \sum_i \left(\frac{\Delta d_i}{d_{i,\text{tol}}} \right)^2$$

can thus be calculated quite easily. Since merely one defect, spherical aberration, is considered, the index i is one. The acceptable fault tolerance amounts to

$$d_{\text{tol}} = \frac{d_{\max} - d_{\min}}{2} = \frac{0.004 - 0.002}{2} = 0.001.$$

The target value is unknown, but can be derived from equation (6.5). It accounts for

$$d_{\text{t}} = S_{\text{I}} = D \cdot n \cdot u' = 0.014 \cdot 1 \cdot 0.35 = 0.0049$$

where the spot diameter D is given by the lateral pixel size l_{pixel}. The relative defect is

$$d_{\text{rel}} = \left| \frac{d_{\text{a}} - d_{\text{t}}}{d_{\text{tol}}} \right| = \left| \frac{0.01 - 0.0049}{0.001} \right| = 5.1.$$

Finally, the wanted absolute value of the merit function amounts to

$$\text{MF} = \sum_i d_{i,\text{rel}}^2 = d_{\text{rel}}^2 = 5.1^2 = 26.01.$$

E25

This exercise is quite comparable to Exercise E24. The given values are
- object angle $u = -0.539°$ (= image angle u', see figure A17) $= -0.0094$ rad,
- $n = 1$,
- $l_{pixel} = 1.7$ μm $= 0.0017$ mm,
- $d_{max} = 0.000\ 04$,
- $d_{min} = 0.000\ 01$, and
- $d_a = -0.0006$ (= the actual Seidel sum S_{II} for coma, see figure A20).

The target defect d_t is given by Seidel sum S_{II} for the considered coma. It follows from the maximum spot dimension (= the lateral pixel size l_{pixel}) that is quantified by tangential coma δ_{tan} according to equation (6.7):

$$d_t = S_{II} = \frac{3}{2} \cdot l_{pixel} \cdot n \cdot u = \frac{3}{2} \cdot 0.0017 \cdot 1 \cdot (-0.0094) = -0.000011.$$

The acceptable fault tolerance is

$$d_{tol} = \frac{d_{max} - d_{min}}{2} = \frac{0.00004 - 0.00001}{2} = 0.000015$$

and the relative defect accounts for

$$d_{rel} = \left| \frac{d_a - d_t}{d_{tol}} \right| = \left| \frac{(-0.0006) - (-0.000011)}{0.000015} \right| = 39.27.$$

The absolute value of the merit function thus accounts for

$$MF = \sum_i d_{i,rel}^2 = d_{rel}^2 = 39.27^2 = 1542.$$

E26

a. In WinLens, the general conditions are defined in the System Parameter Editor as shown in figure A26. A doublet can be chosen from the catalogue database. Here, the available optical components are browsed by using the Query Database function where the range of focal length (and lens diameter) can be defined as shown in figure A27. Since no doublet with the desired focal length is available, the component with the least deviation (here: EFL = 400 mm) is chosen. Its radii of curvature can then be varied manually in order to adjust the focal length, see figure A28. Moreover, the lens diameter can be adjusted (from 31.5 mm to e.g. 42 mm) since the required stop diameter is 40 mm. For such modification, the component has to be unlocked using the convert button at the bottom of the System Data Editor as shown in figure A29.

b. The wave band and wavelengths, respectively, can be defined in the System Parameter Editor, see figure A29. The centre wavelength of the wave band

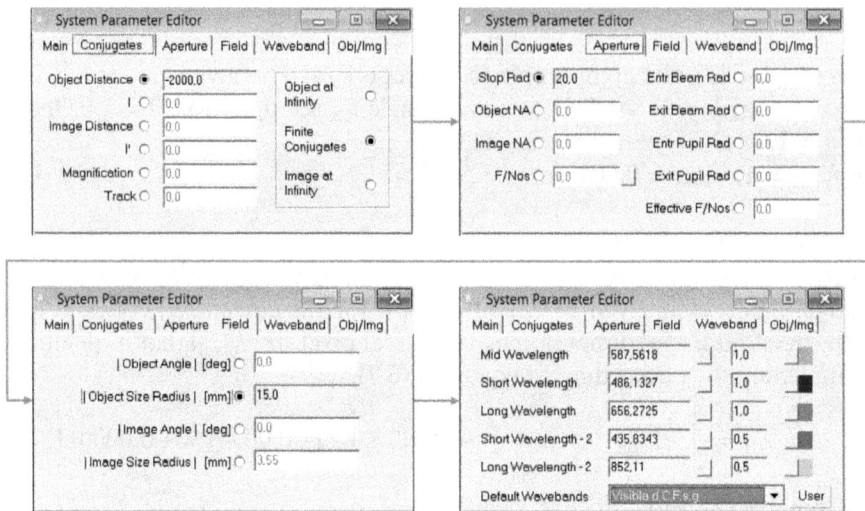

Figure A26. Definition of general conditions in the System Parameter Editor.

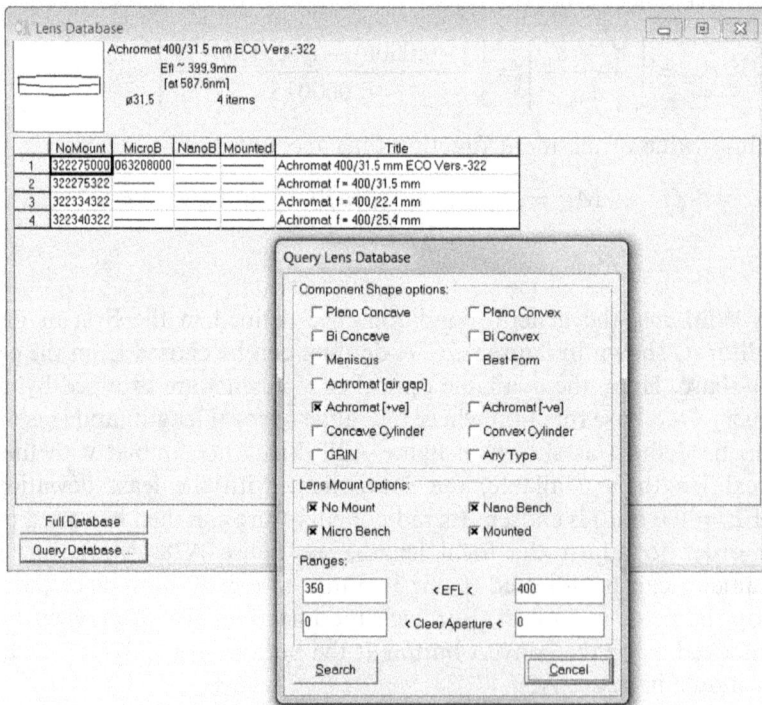

Figure A27. Browsing the catalogue database using the Query Database function.

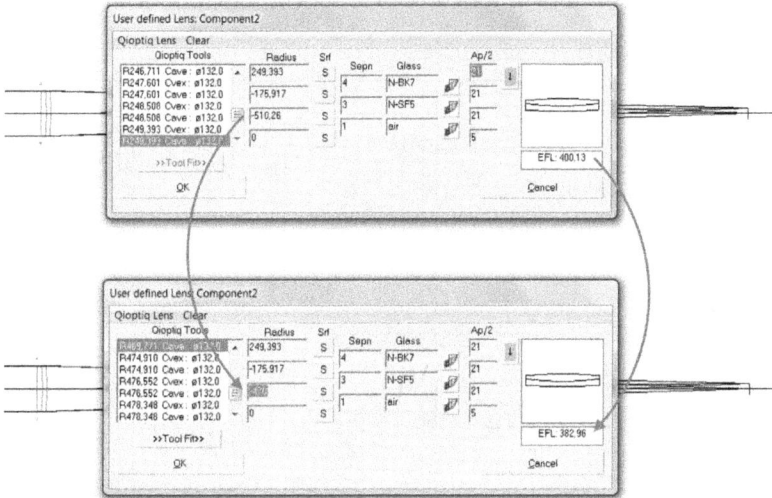

Figure A28. Adjusting the focal length by manual variation of the initial radii of curvature of the catalogue element.

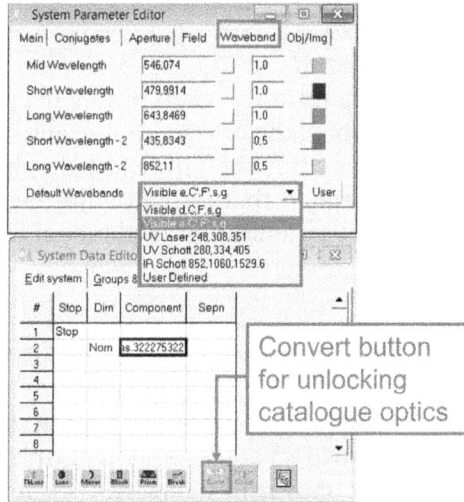

Figure A29. Changing the wave band in the System Parameter Editor (top) as well as unlocking of catalogue optics in order to obtain editable optical elements (bottom).

consisting of the Fraunhofer lines e, C', F', s and g amounts to 546 nm, the wavelength of interest. The evaluation of the resolution can be performed by the spot diagram with an overlaid Airy disc. As shown in figure A30, the system is not diffraction-limited since the image spot is larger than the Airy disc diameter. But, the system can easily be turned into a diffraction-limited system by changing the defocus, i.e. the distance between the doublet and the image plane. At the bottom left of the user interface, the slider 'defocus' is

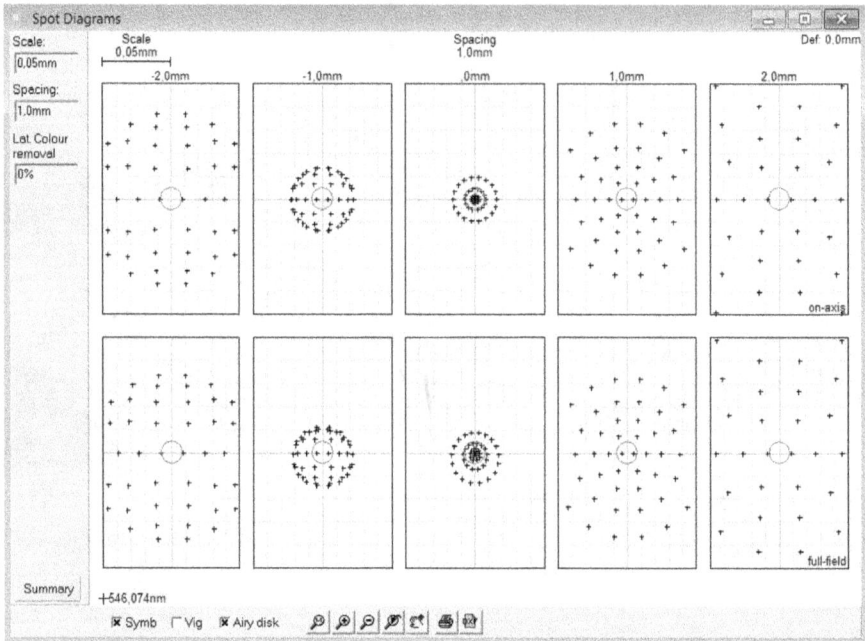

Figure A30. Evaluation of the optical resolution by comparing the spot diameter to the Airy disc (red circle).

pre-defined[8]. When using this slider and shifting the image plane by about 300 μm towards the doublet, the on-axis spot reaches the diffraction limit. This limit is even obtained for the full field by a shift of approximately 410 μm. It can thus be stated that a slight adjustment of the detector or lens position leads to an optimisation of the doublet in terms of optical resolution.

E27

a. The focal length of the lens is 100.16 mm as calculated by WinLens after defining a thick lens with the given lens parameters[9,10], see figure A31(a).

b. The longitudinal chromatic aberration is given by the difference in effective focal length at 643.85 and 479.99 nm displayed in the chromatic aberration diagram. As shown in figure A31(b), it amounts to 101.442 −98.805 = 2.637 mm.

---

[8] This slider can be modified and the other sliders can be customised easily.

[9] Note that there is another glass that is also called SF5, but produced by another manufacturer (CDGM). This glass has another index of refraction than the one produced by Schott. The use of this glass thus leads to another effective focal length of the lens.

[10] If you have chosen the right glass but another effective focal length than 100.16 mm, check for the defined wave band. In the exercise, a centre wavelength of 546 nm = Fraunhofer line *e* is required. For the other visible wave band around the centre wavelength 587 nm (Fraunhofer line *d*), the effective focal length is 100.78 instead of 100.16 mm.

a)

b)

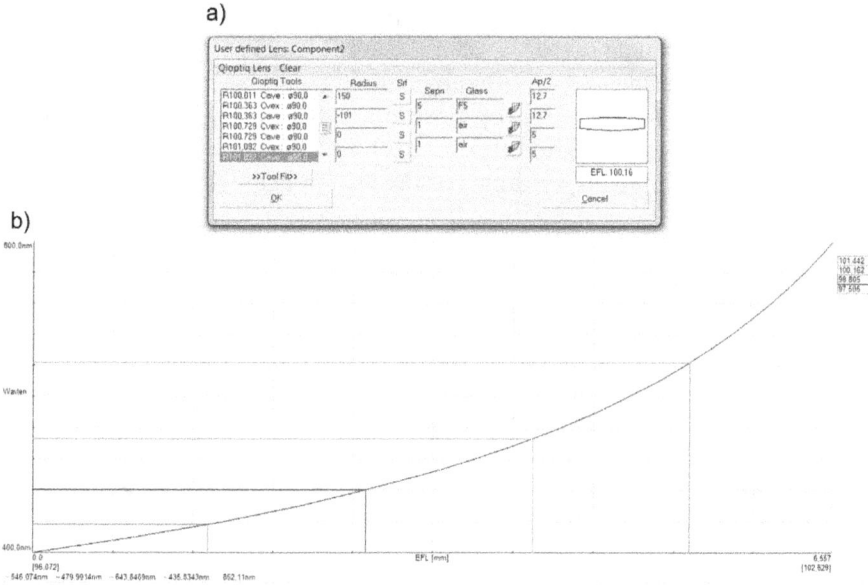

Figure A31. Created lens (a) and its corresponding chromatic aberration diagram (b).

c. A reduction of the longitudinal chromatic aberration without a significant change in effective focal length can be achieved by the use of another glass with almost the same index of refraction, but a higher V-number and lower dispersion, respectively, than the current one. For instance, such a glass can be found in the $n–V$ diagram or using the glass database in Winlens when searching for glasses with a quite similar index of refraction and comparing the V-numbers of the suggested glass. The current glass, F5 from Schott, features an index of refraction of 1.6072 and a V-number of 37.77. The index of refraction and V-number of the crown glass N-SK14 from Schott is 1.6055 and 60.34, respectively. It thus has nearly the same index of refraction, but a much lower dispersion. When replacing F5 by this glass, the effective focal length of the lens amounts to 100.44 mm. The difference is 280 μm and is thus smaller than the given limit of 300 μm. The longitudinal chromatic aberration is $101.261 - 99.604 = 1.657$ mm as determined via the chromatic aberration diagram. This value is notably smaller than before changing the glass. It can thus be stated that this action led to a decrease in chromatic aberration without any significant change in effective focal length—the lens was thus optimised without changing the lens geometry.

E28

a. The Seidel sum for spherical aberration of the given lens and at the given general conditions amounts to $S_I = 2.068\ 60$ at an aperture stop radius of 14 mm.

b. After reducing the stop radius by merely 1 mm ($\approx 7\%$), the Seidel sum is $S_I = 1.537\ 93$; the difference with respect to (a) is thus $\Delta S_I = 0.530\ 67$.

c. The Seidel sum for coma is $S_{II} = 0.1510$ at an aperture stop radius of 14 mm and an object angle of 10°.

d. The stop position can be shifted by defining a distance or air gap thickness between the lens and the stop position in the System Data Editor. After shifting the stop position by 5 mm towards the object, the Seidel sum is $S_{II} = 0.0207$, corresponding to the difference with respect to (c) of $\Delta S_{II} = 0.1303$.

This example shows that an optical system can sometimes be realised easily by a marginal shadowing of incident rays and/or a slight shift of involved optical or mechanical components.

E29

a. The effective focal length of the created achromatic doublet is 100.19 mm. Note that there is no air gap between both lenses since the doublet is cemented according to the given parameters. As shown in figure A32, the second surface of the first lens thus corresponds to the first surface of the second one.

b. The distortion diagram reveals that the achromatic doublet features negative percentaged distortion (→ barrel distortion). It can further be seen that distortion is quite comparable for all considered wavelengths, see figure A32. The Seidel sum for distortion is $S_V = -0.0307$.

c. After shifting the stop position by 2 mm towards the image space[11], i.e. by a value of −2 mm, distortion nearly disappears (see figure A33). The corresponding Seidel sum is reduced from −0.0307 to −0.0037, i.e. by a factor of 8.3.

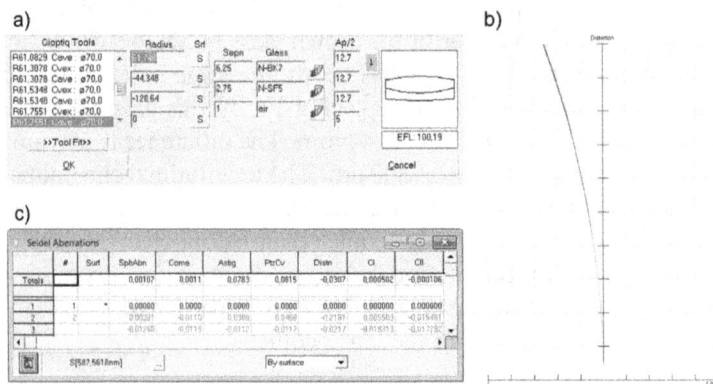

Figure A32. Created lens doublet (a) including the corresponding field diagram for distortion (b) and Seidel aberrations table (c).

---

[11] This shift corresponds to a location of the aperture stop within the material of the first lens. In practice, this can be achieved by realising a virtual pupil.

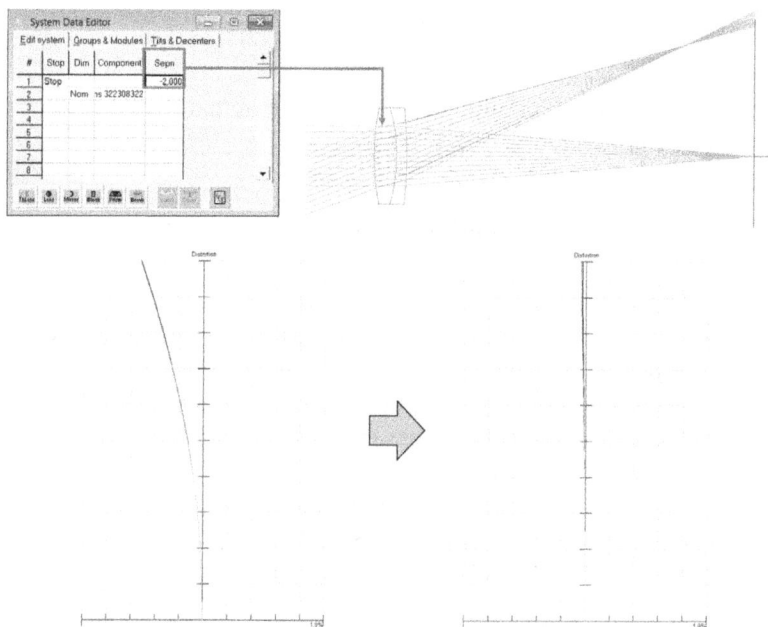

Figure A33. Shifting the stop position towards the image space (top) leads to a notable decrease in distortion (bottom).

d. For the single defect distortion, the values of the merit function are

$$\text{MF} = d_{\text{rel}}^2 = \left(\frac{d_a - d_t}{d_{\text{tol}}}\right)^2 = \left(\frac{(-0.0307) - 0.001}{0.0005}\right)^2 = 4019$$

before and

$$\text{MF} = d_{\text{rel}}^2 = \left(\frac{d_a - d_t}{d_{\text{tol}}}\right)^2 = \left(\frac{(-0.0037) - 0.001}{0.0005}\right)^2 = 88$$

after shifting the stop position. This corresponds to a reduction by a factor of 46.

e. The maximum deviation of image point coordinates from the desired position $D_{\max}$, i.e. the maximum distortion, is given by

$$D_{\max} = \frac{S_V}{2 \cdot n \cdot u}.$$

In the present case, the object angle is $u = 20° = 0.36$ rad. We thus get

$$D_{\max} = \frac{-0.0307}{2 \cdot 1 \cdot 0.36} = -0.043$$

and

$$D_{\max} = \frac{-0.0037}{2 \cdot 1 \cdot 0.36} = -0.005,$$

respectively. The maximum deviation in the actual image height was thus reduced by a factor of 8.6.

A.1.4 Topic: simulation of manufacturing errors and tolerances

E30

In WinLens, a lens surface is usually assumed to be spherical. However, also aspheres, cylinders and toroids can be simulated as shown in figure A34. Using the cylinder/toroid function, a second radius of curvature can be defined (here: 25 mm). For a toroid, this radius is perpendicular to the nominal one (here: 50 mm) and the lens surface consequently consists of two sections, the sagittal and the meridional, with different bending.

After setting the aperture diameter to 30 mm and placing the object at infinity, the spot diagram features elliptical image points that are rotated by 90° to each other, see figure A35. Here, both the scale and spacing were increased in order to visualise this effect. This behaviour clearly indicates the formation of astigmatism even though the incident light is not inclined with respect to the optical axis (since the object is placed at infinity). For a scaling and placing of 10 mm, also the approximate position of the circle of least confusion can be identified.

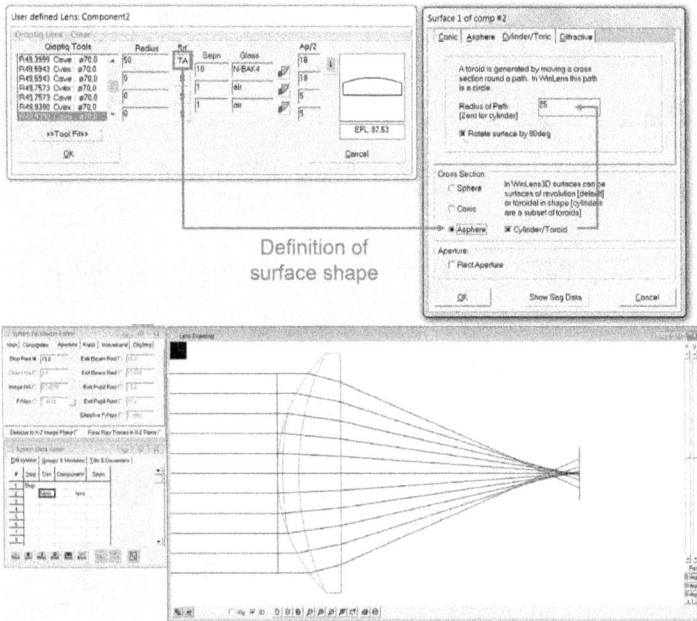

Figure A34. Definition of a toroid on the lens surface (top) including a drawing of such a lens (bottom).

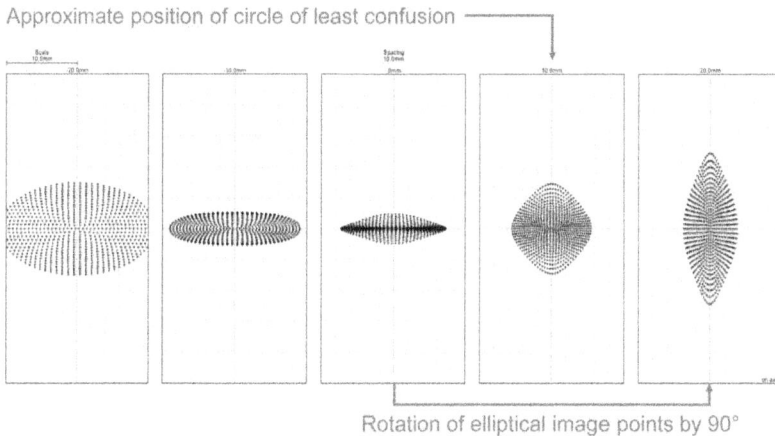

Figure A35. Visualisation of the formation of astigmatism at toric lens surfaces.

E31

  a. The peak-to-valley value of the transmitted wave front is PV = 8.05 waves at a wavelength of 546 nm.
  b. After tilting the lens by 5°, the peak-to-valley value is 21.8 waves, corresponding to an increase by a factor of 2.7.
  c. An additional decentre of the lens by 5 mm results in a peak-to-valley value of 37.3 waves. With respect to the perfectly-aligned lens without any position error, this corresponds to an increase by a factor of 4.6.

All three cases are summarised in figure A36.

E32

The created lens is a biconvex lens with an effective focal length of 14.69 mm. By the tilt of a second lens surface by 10° the PV and rms values are notably increased by a factor of 5.3 and 3.1, respectively, see figure A37.

E33

  a. The comparison of both cases via the spot diagram and the wave front plot is shown in figure A38. It can be seen that (i) the focal point or image spot is much smaller and closer to the Gaussian image space for the curved surface being orientated towards the object and (ii) the wave front deformation is much higher for the first case (plane surface towards object space). Here, the peak-to-valley (PV) value is 20.6 waves = 20.6 × 546.074 nm ≈ 11.249 μm and the root-mean-square (rms) value is 13 waves ≈ 7.098 μm. If the curved surface is orientated towards the object space—using the 'Rev' (= 'reverse') button in the column 'Dirn' (= 'direction') of the System Data Editor—the PV value is reduced by a factor of about 3.8 (from 20.6 to 5.32 waves) and the rms value decreases by quite the same factor (from 13 to 3.36 waves). We

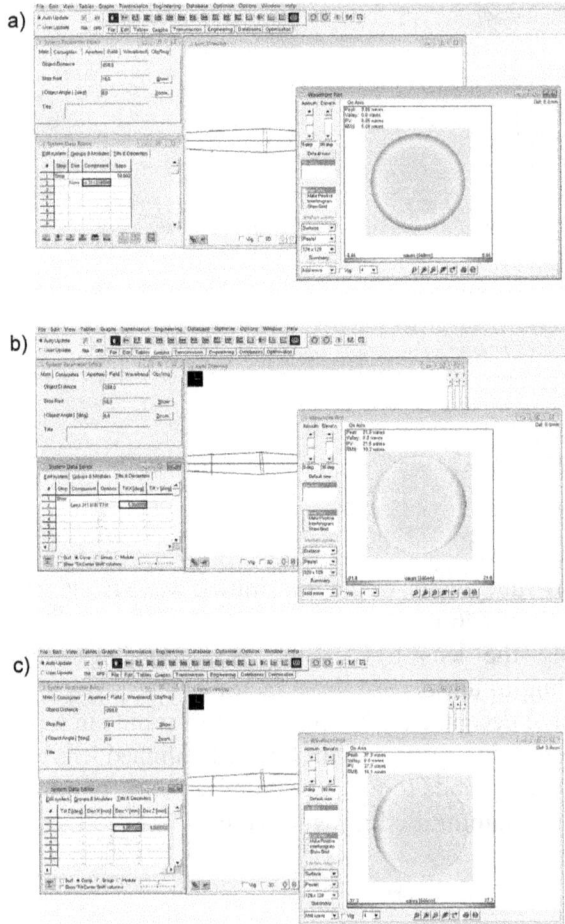

Figure A36. Comparison of the lens orientation and position as well as the wave front plot for (a) perfect alignment, (b) tilt by 5°, and (c) tilt by 5° and additional decentre by 5 mm.

see that the lens orientation notably affects the imaging quality. This point is of specific importance during the mounting of any opto-mechanical system.

b. If the first surface is tilted from 0° to 0.1°, the PV value slightly increases by 0.2 waves (from to 20.6 to 20.8 waves) if the plane surface is orientated towards the object. For the other configuration where the curved side is orientated towards the object a similar change in PV value of 0.19 waves (from 5.32 waves 5.51 waves) occurs.

c. A tilt of the entire lens from 0° to 0.1° results in an increase in PV value of 0.4 waves for the lens being orientated with its plane surface towards the object space. For the other orientation the PV value increases by 0.06 waves.

d. The decentre results in an increase in PV value by 0.8 waves if the plane surface is orientated towards the object. For the other orientation (curved surface towards object space), the PV value is increased by 0.22 waves due to

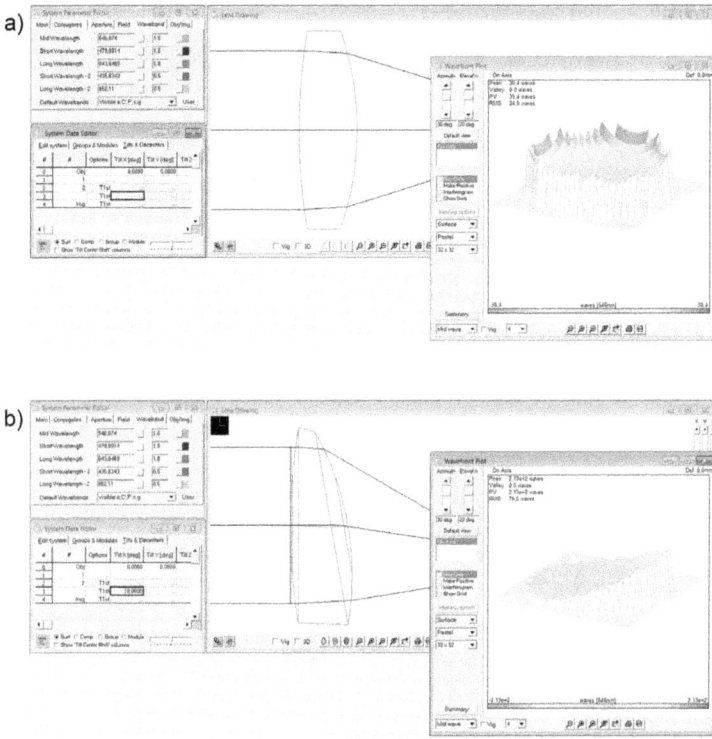

Figure A37. Comparison of the 3D wave front plot of a perfect lens (a) and a lens with a tilted back surface (b).

the decentre. The lens orientation has a notable impact on wave front distortion. In the case of an orientation of the plane surface towards the object, a higher PV value is found.

e. A decentre in the z-direction corresponds to a defocus of the image plane. For the plane lens surface being orientated towards the incoming light, we get a difference in PV value by 0.9. The same value is found if the curved lens surface is orientated towards the object.

f. When comparing the particular values as listed in table A1 it turns out that a distance error, i.e. a defocus, has the highest impact on the PV value.

E34

The doublet can be defined by three surfaces of a user-defined lens as shown in figure A39. It is then a single component. Otherwise, it could not be tilted as an entire component.

a. Orientating the lens surface with the lower radius of curvature (i.e. the surface with the higher bending) towards the incoming light leads to a notable reduction in wave front distortion and spot size as shown by the comparison in figure A40. In the latter configuration, the lens is even

Plane surface towards object:　　　　Curved surface towards object:

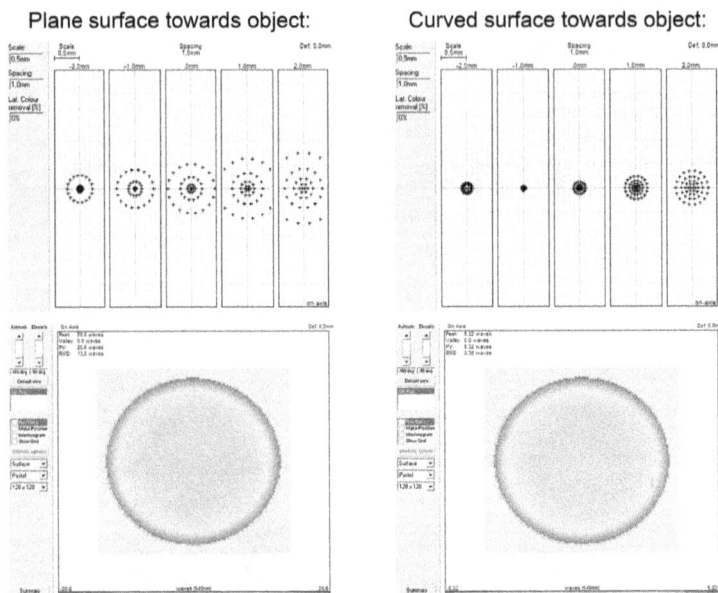

Figure A38. Comparison of the spot diagram and the 3D wave front plot for a plano-convex lens at different orientations.

Table A1. Comparison of lens orientation and position errors and accompanying change in PV value.

Orientation and position error	Change in PV value (waves)
Plane towards object, surface tilt	0.2
Curve towards object, surface tilt	0.19
Plane towards object, lens tilt	0.4
Curve towards object, lens tilt	0.06
Plane towards object, lateral offset	0.8
Curve towards object, lateral offset	0.22
Plane towards object, distance error	0.9
Curve towards object, distance error	0.9

    diffraction-limited in the Gaussian image space as visualised by adding the Airy disc diameter to the spot diagram.

b. As expected, a tilt of a lens surface leads to a reduction in imaging quality. First, the wave front distortion increases from PV = 0.113 waves to 0.301 waves, i.e. by a factor of about three. Further, coma arises as shown by the shape of the image point in the Gaussian image plane (see figure A41).

c. Both lenses can be treated or modified independently from each other when entering two single lenses. If the air gap or distance between both lenses is zero, a kind of cemented doublet is simulated.

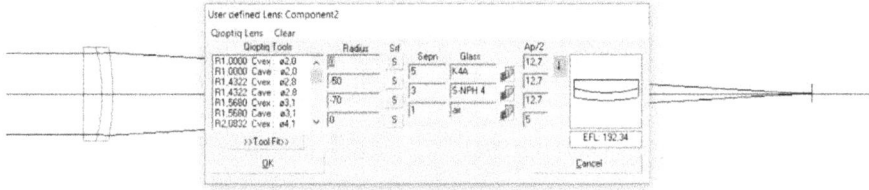

Figure A39. Definition of an achromatic doublet as a single component.

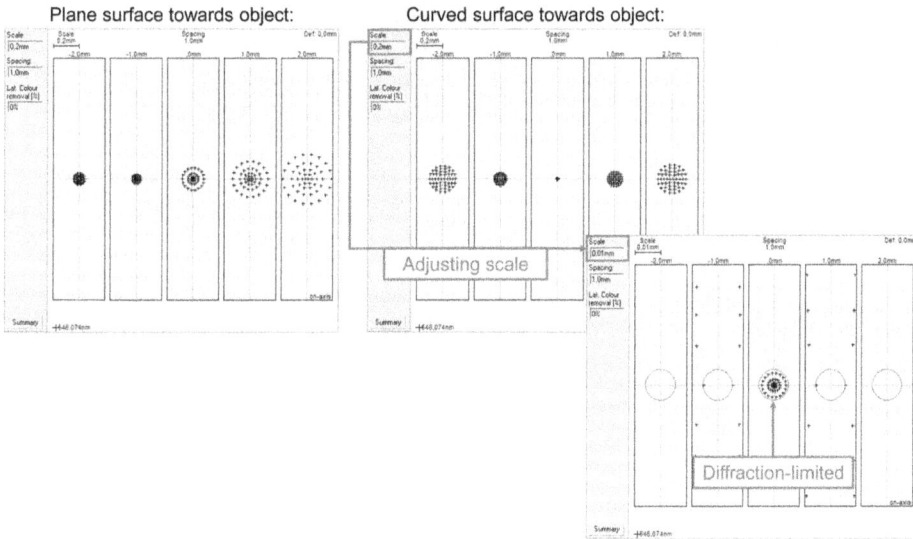

Figure A40. Comparison of the spot diagram of the achromatic doublet at different orientations.

d. For a tilt of the second lens by 0.5° around the *Y*-axis we also observe an increase in wave front distortion (increase in PV value from 1.97 waves to 4.02 waves) and the formation of coma as already simulated in (1b).

e. The specification 3/10 means that a maximum of 10 Newton fringes is allowed to be the deviation of the lens surface from its target value. Quantitatively, one fringe is half the test wavelength. The maximum deviation of a single surface is thus 10×633 nm/2 = 5×633 nm = 3.165 µm. The maximum error that can arise if both surfaces feature the maximum deviation of 3.165 µm is finally 6.33 µm since the specified tolerance may be positive for one surface and negative for the other one. For instance, the radius of curvature of the second surface of the first lens can be 49.997 and 50.003 mm for the first surface of the second lens. If we consider this fact and modify the 'perfect' lenses considering the possible radii of curvature calculated above we get a significant decrease in image spot size as shown in figure A42. It turns out that here, a manufacturing error within specification can even be advantageous! In other words, the initial design of this doublet was not the best solution.

Plane surface towards object: Curved surface towards object:

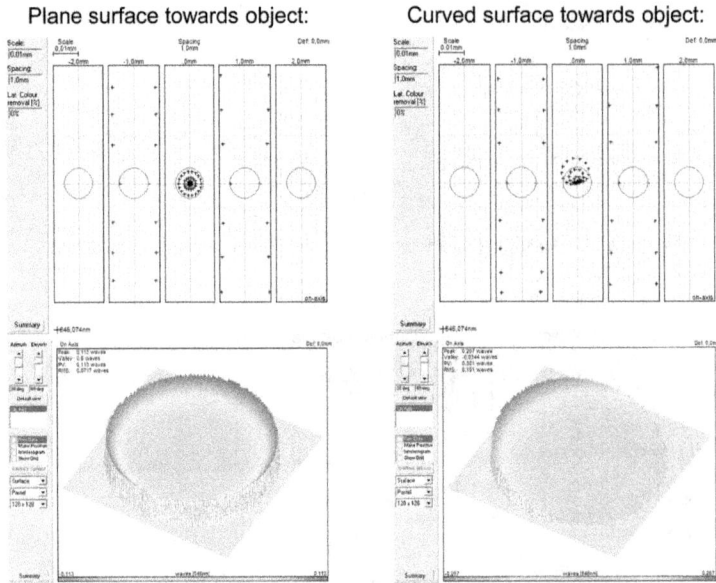

Figure A41. Increase in wave front distortion as well as formation of coma due to a tilt of single surfaces of the achromatic doublet.

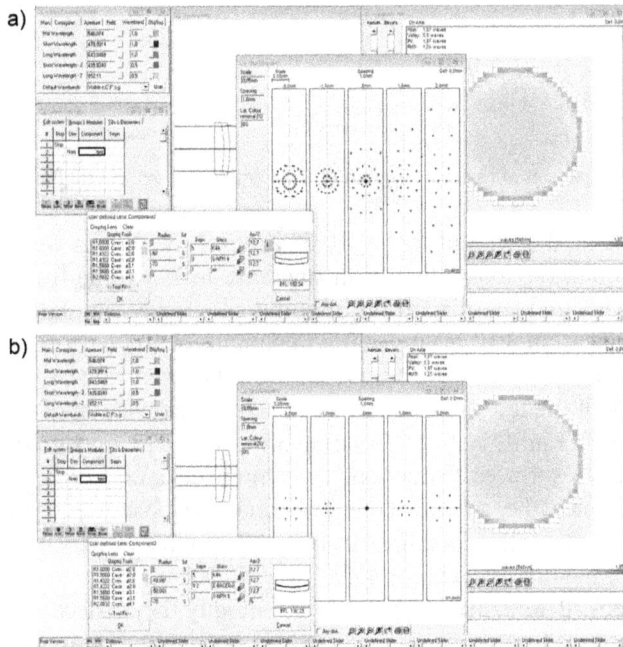

Figure A42. Decrease in image spot size by applying maximum acceptable manufacturing tolerances of radii of curvature.

A.2 Formulary

Here, the most important selected equations for the design of lenses and optical systems are summarised.

Snells's law:

$$n_1 \cdot \sin \varepsilon_1 = n_2 \cdot \sin \varepsilon_2$$

n = index of refraction in front of (index 1) and behind (index 2) interface, ε_1 angle of incidence, ε_2 angle of refraction.

V-number:

$$V_e = \frac{n_e - 1}{n_{F'} - n_{C'}}$$

n = index of refraction (at Fraunhofer lines e, F', and C').

Airy disc diameter D_{Airy}:

$$D_{\text{Airy}} = 1.22 \cdot \frac{\lambda \cdot f}{D_{\text{stop}}}$$

λ = wavelength of light, f = is focal length, D_{stop} = stop diameter.

Rayleigh-criterion:

$$d_{\min} = \frac{D_{\text{Airy}}}{2}$$

$d_{\min}$ = minimum distance of two neighboured Airy discs.

Lensmaker's equation for thin lenses:

$$\frac{1}{\text{EFL}} = (n_l - 1) \cdot \left(\frac{1}{R_1} - \frac{1}{R_2} \right)$$

EFL = effective focal length, n_l = index of refraction of lens material, R = radius of curvature.

Lensmaker's equation for thick lenses:

$$\frac{1}{\text{EFL}} = (n_l - 1) \cdot \left(\frac{1}{R_1} - \frac{1}{R_2} + \frac{(n_l - 1) \cdot t_c}{n_l \cdot R_1 \cdot R_2} \right)$$

EFL = effective focal length, n_l = index of refraction of lens material, R = radius of curvature, t_c = centre thickness.

Imaging equation:

$$\frac{1}{\text{EFL}} = \frac{1}{a} - \frac{1}{a'}$$

EFL = effective focal length, a = object distance, a' = image distance.

Magnification m:

$$m = \frac{a'}{a} = \frac{y'}{y}$$

a' = image distance, a = object distance, y' = image height, y = object height.

Total effective focal length EFL_{tot} *of lens doublet*:

$$\frac{1}{EFL_{tot}} = \frac{1}{EFL_1} + \frac{1}{EFL_2} - \frac{d}{EFL_1 \cdot EFL_2}$$

EFL = effective focal length, d = distance between involved lenses.

Condition for achromatism:

$$EFL_1 \cdot V_1 = -EFL_2 \cdot V_2$$

EFL = effective focal length, V = V-number.

Seidel coefficient A:

$$A = n \cdot \left(h \cdot \frac{1}{R} + u \right)$$

n = index of refraction, h = ray entrance height, R = radius of curvature, u = aperture angle.

Nominal back focal length BFL of a curved optical interface:

$$BFL = R \cdot \frac{n}{n-1}$$

R = radius of curvature, n is the index of refraction behind interface.

Seidel sum for spherical aberration S_I:

$$S_I = -\sum_{i=1}^{k} A_i^2 \cdot h_i \cdot \Delta_i \left(\frac{u}{n} \right)$$

with

$$A_i = n_i \cdot \left(h_i \cdot \frac{1}{R_i} + u_i \right)$$

and

$$\Delta_i \left(\frac{u}{n} \right) = \frac{u_{i+1}}{n_{i+1}} - \frac{u_i}{n_i}$$

Seidel sum for coma S_{II}:

$$S_{II} = -\sum_{i=1}^{k} A_i \cdot \bar{A}_i \cdot h_i \cdot \Delta_i \left(\frac{u}{n} \right)$$

Seidel sum for astigmatism S_{III}:

$$S_{\text{III}} = -\sum_{i=1}^{k} \bar{A}_i^2 \cdot h_i \cdot \Delta_i\left(\frac{u}{n}\right)$$

Petzval radius R_{Petzval}:

$$R_{\text{Petzval}} = \frac{H^2}{n \cdot S_{\text{IV}}}$$

n = the index of refraction behind interface, S_{IV} = fourth Seidel sum, H = system-specific parameter given by

$$H = n \cdot (u \cdot \bar{h} - \bar{u} \cdot h)$$

Seidel sum for Petzval field curvature S_{IV}:

$$S_{\text{IV}} = -\sum_{i=1}^{k} H_i^2 \cdot \frac{1}{R_i} \cdot \Delta_i\left(\frac{u}{n}\right)$$

with

$$\Delta_i\left(\frac{1}{n}\right) = \frac{1}{n_{i+1}} - \frac{1}{n_i}$$

Distortion D:

$$D = y'_a - y'_p$$

y'_a = actual image point coordinate, y'_p = paraxial image point coordinate.
   Percentaged distortion $D_\%$:

$$D_\% = \frac{y'_a - y'_p}{y'_p} \cdot 100\%$$

y'_a = actual image point coordinate, y'_p = paraxial image point coordinate.
   Seidel sum for distortion S_{V}:

$$S_{\text{V}} = -\sum_{i=1}^{k} \left[\frac{\bar{A}_i^3}{A_i} \cdot h_i \cdot \Delta_i\left(\frac{u}{n}\right) + \frac{\bar{A}_i}{A_i} \cdot H_i^2 \cdot \frac{1}{R_i} \cdot \Delta_i\left(\frac{1}{n}\right)\right]$$

or

$$S_{\text{V}} = -\sum_{i=1}^{k} \frac{\bar{A}_i}{A_i} \cdot [(S_{\text{III}})_i + (S_{\text{IV}})_i]$$

Michelson contrast C_{M}, *a.k.a. modulation M*:

$$C_{\text{M}} = M = \frac{I_{\max} - I_{\min}}{I_{\max} + I_{\min}}$$

$I_{\max}$ = light intensity within bright object areas, $I_{\min}$ = light intensity at dark object areas.

Modulation transfer function MTF:

$$\text{MTF} = \frac{M_{\text{image}}}{M_{\text{object}}}$$

M = modulation or (contrast).

Cut-off-frequency $f_{\text{cut-off}}$:

$$f_{\text{cut-off}} = \frac{1}{\arctan\left(\dfrac{\lambda}{D}\right)}$$

λ = wavelength, D = diameter of entrance pupil.

F-number:

$$\text{f-number} = \frac{\text{EFL}}{D_{\text{aperture}}}$$

EFL = effective focal length, D_{aperture} = aperture diameter.

Merit function MF:

$$\text{MF} = \sum_i d_{i,\text{rel}}^2 = \sum_i \left(\frac{d_{i,\text{a}} - d_{i,\text{t}}}{d_{i,\text{tol}}}\right)^2 = \sum_i \left(\frac{\Delta d_i}{d_{i,\text{tol}}}\right)^2$$

with

$$d_{\text{tol}} = \frac{d_{\text{max}} - d_{\text{min}}}{2}$$

D = defect (indices: rel = relative, a = actual, t = target, tol = tolerance).

Image spot diameter D:

$$D = \frac{S_{\text{I}}}{n \cdot u'}$$

S_{I} = Seidel sum for spherical aberration, n = index of refraction of surrounding medium, u' = image angle.

Defocus φ:

$$\varphi = \frac{3}{8} \cdot \frac{S_{\text{I}}}{n \cdot u^2}$$

S_{I} = Seidel sum for spherical aberration, n = index of refraction of surrounding medium, u = object angle.

Tangential coma δ_{tan}:

$$\delta_{\text{tan}} = \frac{3}{2} \cdot \frac{S_{\text{II}}}{n \cdot u}$$

S_{II} = Seidel sum for coma, n = index of refraction of surrounding medium, u = object angle.

Sagittal coma δ_{tan}:

$$\delta_{\mathrm{sag}} = \frac{1}{2} \cdot \frac{S_{\mathrm{II}}}{n \cdot u}$$

S_{II} = Seidel sum for coma, n = index of refraction of surrounding medium, u = object angle.

Maximum distance between outermost image section and Gaussian image plane φ_{max}:

$$\varphi_{\mathrm{max}} = -\frac{S_{\mathrm{III}}}{n \cdot u^2}$$

S_{III} = Seidel sum for astigmatism, n = index of refraction of surrounding medium, u = object angle.

Maximum deviation of image point coordinates from desired position D_{max}:

$$D_{\mathrm{max}} = \frac{S_{\mathrm{V}}}{2 \cdot n \cdot u}$$

S_{V} = Seidel sum for distortion, n = index of refraction of surrounding medium, u = object angle.

Numerical aperture NA:

$$\mathrm{NA} = n_s \cdot \sin(u)$$

n_s = index of refraction of surrounding medium, u = aperture angle.

Wave front distortion Δw:

$$\Delta W = \Delta n \cdot 2 \cdot t$$

Δn = deviation in index of refraction, t = thickness.

A.3 Further reading

A.3.1 Books

- Bentley J and Olson C 2012 *Field Guide to Lens Design* 1st edn (Bellingham, WA: SPIE Press) ISBN 9780819491640
- Fischer R E 2008 *Optical System Design* 1st edn (New York: McGraw-Hill) ISBN 978-0071472487
- Geary J M 2002 *Introduction to Lens Design with Practical Zemax Examples* 1st edn (Richmond, VA: Willmann-Bell) ISBN 978-0943396750
- Gerhard C 2017 *Optics Manufacturing: Components and Systems* 1st edn (Boca Raton, FL: CRC Taylor & Francis) ISBN 978-1-4987-6459-9
- Gerhard C 2016 *Tutorium Optik* 1st edn (Heidelberg: Springer Verlag) ISBN 978-3-662-48574-3 (in German)
- Kingslake R and Johnson R B 2010 *Lens Design Fundamentals* 2nd edn (Cambridge, MA: Academic Press) ISBN 978-0123743015
- Kingslake R 1983 *Optical System Design* 1st edn (Cambridge, MA: Academic Press) ISBN 9780323141109

- Malacara D 2006 *Optical Shop Testing* 3rd edn (Hoboken, NJ: Wiley-Interscience) ISBN 9780471484042
- Smith W J 1992 *Modern Lens Design* 1st edn (New York: McGraw-Hill) ISBN 978-1259583759
- Smith W J 2007 *Modern Optical Engineering* 4th edn (New York: McGraw-Hill Professional) ISBN 978-0071476874
- Sun H 2017 *Lens Design: A Practical Guide* 1st edn (Boca Raton, FL: CRC Press) ISBN 978-1138455702
- Walther A 2006 *The Ray and Wave Theory of Lenses* 1st edn (Cambridge: Cambridge University Press) ISBN 978-0521028295
- Welford W 1986 *Aberrations of Optical Systems* 1st edn (Boca Raton, FL: Taylor & Francis) ISBN 978-0852745649
- Yoder P 2008 *Mounting Optics in Optical Instruments* 2nd edn (Bellingham, WA: SPIE Publications) ISBN 978-0819471291
- Yoder P 2005 *Opto-Mechanical Systems Design* 3rd edn (Boca Raton, FL: CRC Press) ISBN 978-1574446999.

A.3.2 Articles and conference proceedings

Actually, there are a notable number of papers on optical system design and all related fields. The following articles and conference proceedings were selected since here, the same software as used in the present book—WinLens and PreDesigner—was applied. Most of the papers are freely available on the particular publisher's website.

- Adams G, Thöniß T and Gerhard C 2013 Designed to disperse: easy modelling of prism and grating spectrometers and more *Opt. Photonik* **8** 50–3
- Gerhard C and Adams G 2010 Correction of chromatic aberration - from design to completed lens systems *Imaging Microsc.* **3** 39–40
- Gerhard C and Adams G 2015 Easy-to-use software tools for teaching the basics, design and applications of optical components and systems *Proc. SPIE* **9793** 97930N
- Gerhard C and Adams G 2013 Optical design with WinLens™3D – Part 1: preliminary viewing with PreDesigner *optolines* **31** 4–5
- Gerhard C and Adams G 2013 Optical design with WinLens™3D – Part 2: simulation and analysis *optolines* **32** 12–13
- Gerhard C and Adams G 2013 Optical design with WinLens™3D – Part 3: analysis and optimisation *optolines* **33** 11–13
- Gerhard C, Adams G and Wienecke S 2010 Design and manufacture of achromatic lenses *LED Prof. Rev.* **19** 40–3
- Harendt N and Gerhard C 2008 Simulation and optimization of optical systems *LED Prof. Rev.* **10** 40–2
- Schuhmann R G and Adams G 2001 Enhancements to the optimization process in lens design (I) *Proc. SPIE* **4441** 30

- Schuhmann R G and Adams G 2001 Enhancements to the optimization process in lens design (II) *Proc. SPIE* **4441** 37
- Schuhmann R G and Adams G 2004 Active glass maps for an optical design program *Proc. SPIE* **5249**, *Opt. Des. Eng.* **364**, doi:10.1117/12.517142
- Thöniß T, Adams G and Gerhard C 2009 Optical system design – software tools cover envelope calculations to the final engineering drawings *Opt. Photonik* 30–3.

www.ingramcontent.com/pod-product-compliance
Lightning Source LLC
Chambersburg PA
CBHW080549220326
41599CB00032B/6417